MEXICAN STUDIES IN THE HISTORY AND PHILOSOPHY OF SCIENCE

BOSTON STUDIES IN THE PHILOSOPHY OF SCIENCE

VOLUME 172

MEXICAN STUDIES IN THE HISTORY AND PHILOSOPHY OF SCIENCE

Edited by

SANTIAGO RAMIREZ
Universidad National Autónoma de México

and

ROBERT S. COHEN
Boston University

KLUWER ACADEMIC PUBLISHERS
DORDRECHT / BOSTON / LONDON

Library of Congress Cataloging-in-Publication Data

Mexican studies in the history and philosophy of science / edited by Santiago Ramirez and Robert S. Cohen.
p. cm. -- (Boston studies in the philosophy of science ; 172)
Includes index.
ISBN-13: 978-94-010-6535-1 (alk. paper)
1. Science--Mexico--History. 2. Science--Mexico--Philosophy.
I. Ramírez, Santiago. II. Cohen, R. S. (Robert Sonné) III. Series.
Q174.B67 vol. 172
[Q127.M]
001'.01 s--dc20
[500] 95-13783

ISBN-13: 978-94-010-6535-1 e-ISBN-13: 978-94-009-0109-4
DOI: 10.1007/978-94-009-0109-4

Published by Kluwer Academic Publishers,
P.O. Box 17, 3300 AA Dordrecht, The Netherlands.

Kluwer Academic Publishers incorporates
the publishing programmes of
D. Reidel, Martinus Nijhoff, Dr W. Junk and MTP Press.

Sold and distributed in the U.S.A. and Canada
by Kluwer Academic Publishers,
101 Philip Drive, Norwell, MA 02061, U.S.A.

In all other countries, sold and distributed
by Kluwer Academic Publishers Group,
P.O. Box 322, 3300 AH Dordrecht, The Netherlands.

Printed on acid-free paper

Softcover reprint of the hardcover 1st edition 1995

TABLE OF CONTENTS

D. MATHEMATICS

PREFACE

For a North American seeking to know the Mexican mind, and especially the sciences today and in their recent development, a great light of genius is to be found in Mexico City in the late 17th century. The genius is that of one who surely may be counted as the first Mexican philosopher of nature, a nun of the Order of Saint Jerome: Sor Juana Inés de la Cruz.

Sor Juana must speak for herself, from her penetrating exercise of an independent mind within a political and religious formation which denigrated women and circumscribed reason itself. To understand this world of ours, to join in an enlightenment which would be both natural and inspired, Sor Juana clearly understood the requirements of learning, observing, logic and reasoning.

> In darkness foundering
> Words fail the troubled mind.
> For who, I ask, can light me
> When Reason is blind?

Even now, after the great steps toward liberation of women, and the substantial scientific contributions toward sheer empirical awareness of both the multiple orders of Nature and the subtle aesthetics of individual art and social harmony, we too in the earthly world of the 20th century must affirm what she affirmed. As preface to this volume of contemporary Mexican studies in the history and philosophy of mathematics and the natural and social sciences, I ask readers to listen to Sor Juana,[1] and each to take her 'Holy Theology' for your own goal, your own hope for reason to be found in your world:

> I went on in this way, always directing each step of my studies, as I have said, toward the summit of Holy Theology; but it seemed to me necessary to ascend by the ladder of the humane arts and sciences in order to reach it; for who could fathom the style of the Queen of Sciences without knowing that of her handmaidens? Without Logic, how should I know the general and specific methods by which Holy Scripture is written?

S. Ramirez and R. S. Cohen (eds.), Mexican Studies in the History and Philosophy of Science, vii–ix.

Without Rhetoric, how should I understand its figures, tropes, and locutions? Or how, without Physics or Natural Science, understand all the questions that naturally arise concerning the varied natures of those animals offered in sacrifice, in which a great many things already made manifest are symbolized, and many more besides? How should I know whether Saul's cure at the sound of David's harp was owing to a virtue and power that is natural in Music or owing, instead, to a supernatural power that God saw fit to bestow on David? How without Arithmetic might one understand all those mysterious reckonings of years and days and months and hours and weeks that are found in Daniel and elsewhere, which can be comprehended only by knowing the natures, concordances, and properties of numbers? Without Geometry, how could we take the measure of the Holy Ark of the Covenant or the Holy City of Jerusalem, each of whose mysterious measurements forms a perfect cube uniting their dimensions, and each displaying that most marvelous distribution of the proportions of every part? Without the science of Architecture, how understand the mighty Temple of Solomon – where God Himself was the Draftsman who set forth His arrangement and plan, and the Wise King was but the overseer who carried it out; where there was no foundation without its mystery, nor column without its symbol, nor cornice without its allusion, nor architrave without its meaning, and likewise for every other part, so that even the very least fillet served not only for the support and enhancement of Art, but to symbolize greater things? How, without a thorough knowledge of the order and divisions by which History is composed, is one to understand the Historical Books – as in those summaries, for example, which often postpone in the narration what happened first in fact? How, without command of the two branches of Law, should one understand the Books of Law? Without considerable erudition, how should we understand the great many matters of profane history that are mentioned by Holy Scripture: all the diverse customs of the Gentiles, all their rituals, all their manners of speech? Without knowing many precepts and reading widely in the Fathers of the Church, how could one understand the obscure sayings of the Prophets? Well then, and without being expert in Music,

> how might one understand those musical intervals and their perfections that occur in a great many passages – especially in Abraham's petitions to God on behalf of the Cities, beseeching God to spare them if there were found fifty righteous people within? And the number fifty Abraham reduced to forty-five, which is sesquinonal [10 to 9] or like the interval from mi to re; this in turn he reduced to forty, which is the sesquioctave [9 to 8] or like the interval from re to mi; thence he went down to thirty, which is sesquitertia, or the interval of the diatessaron [the perfect fourth]; thence to twenty, the sesquialtera or the diapente [the fifth]; thence to ten, the duple, which is the diapason [the interval and consonance of the octave]; and because there are no more harmonic intervals, Abraham went no further. How could all this be understood without knowledge of Music? Why, in the very Book of Job, God says to him: "*Shalt thou be able to join together the shining stars the Pleiades, or canst thou stop the turning about of Arcturus? Canst thou bring forth the day star in its time, and make the evening star to rise upon the children of the earth?*" Without knowledge of Astronomy, these terms would be impossible to understand. Nor are these noble sciences alone represented; indeed, not one of the mechanical arts escapes mention. In sum, we see how this Book contains all books, and this Science includes all sciences.

Sor Juana died on April 17, 1695, three centuries ago.

Robert S. Cohen

NOTE

[1] The first short piece is from the translation of Luis Harss *Sor Juana's Dream* (New York: Lumen Books, 1986, p. 1). The longer passage is from *La Respuesta/The Answer* in the translation of Electa Arenal and Amanda Powell, and is used with their kind permission (New York: The Feminist Press at the City University of New York, 1993; pp. 53 and 55). Their volume includes selected poems by Sor Juana and helpful notes. Another translation, also with poems, is that of Alan Trueblood, *A Sor Juana Anthology* (Cambridge: Harvard University Press, 1988). The principal study available in English is that of Octavio Paz, *Sor Juana, or, The Traps of Faith* (tr. Margaret Sayers Peden, Cambridge: Harvard University Press, 1988).

INTRODUCTION

I am uniquely gratified to be able to present and edit, with the most unsparing support of Robert Cohen, the following selection to show the recent state of the philosophy and history of the sciences in Mexico.

It must be said that, in general, the texts chosen do not necessarily relate to the history of Mexican sciences or to an imaginary 'Mexican philosophy'. I have tried not to be constrained by the national and regional issues. This does not mean that they should be dismissed or overlooked. Rather, it is a different way of approaching them. As Octavio Paz puts it in the *Labyrinth of Solitude*,

> We live, as the rest of the planet in a decisive and deadly crossroad, orphans without a past yet with a future to be created. Universal history is now a common task . . . and our labyrinth, the labyrinth of all men.

Mexican history and philosophy of science is not a new matter. From the very early days of colonial domination, Mexican institutions of knowledge harbored scholars who were to discuss the problems of science in a very outstanding manner. Fray Alonso de la Veracruz, for example, is still waiting for the necessary research that would unearth his most ingenious talents. Later, Sor Juana Ines de la Cruz consecrated some of her major works to the subject. Her production has been unveiled as the result of one of the most profound visions of her time and yet she was one of the most tragic victims of the 'traps of faith'.

Aufklärung also found its way into our continent, provoking the Wars of Independence and the turmoil that would last throughout the first half of the 19th century. In the midst of chaos, some of our forefathers found time to speculate on the role of science, its institutions and their history. (Such is the case of Santiago Ramirez as is discussed by José Rüiz de Esparza.)

It was only after the victorious war against the French-inspired 'Empire', in 1868, that a group of Mexican intellectuals – followers of Auguste Comte – consecrated their efforts to promote a new concep-

S. Ramirez and R. S. Cohen (eds.), Mexican Studies in the History and Philosophy of Science, xi–xii.

tion of science and of its role in the reconstruction of a society and a nation. The positivist group, headed by Gabino Barreda, discussed Darwin's theories. (The contents of their arguments are expounded here by Rosaura Ruiz.)

The 1910 Revolution changed everything again. By the 1930's, science was viewed, again, as an instrument for liberation and emancipation. In the 50's, inspired by Marxism, Eli de Gortari organized the milestone *Seminario de Problemas Científicos y Filosóficos* that gathered the most distinguished personalities in the history and philosophy of the sciences and in the sciences themselves.

It can be claimed that De Gortari's influence and moral integrity marked the development of these disciplines: in 1968 traditional Marxism became obsolete. As part of that critique, I offer the work by Santiago Ramirez Ruiz on Psychoanalysis and Marxism, a line of thought developed to try to solve the problems raised by the vacuum left by 'historical materialism'. (Two basic issues around it are discussed by Corina Yturbe and Carlos Pereyra.)

Another line of thought was followed by scientists. One of such, an outstanding example, is the work of Carlos Torres.

Mexican philosophy and history of science can thus be put into perspective. Brought forth by the need to construct a national identity, it searched for inspiration, and too often for its answers, elsewhere: Positivism or Marxism. I think it has finally reached its maturity: it can deal with problems that are not confined to a chauvinist nationalism that would isolate Mexican thought, and also has learned not to repeat what is being done in the metropolis. Mexican history and philosophy of the sciences has arrived, at last, to a point in which, as Borges wanted, "we can claim for ourselves every tradition".

There are other reasons. Subjectivity cannot be excluded. I have chosen, thus, to include the papers by Santiago Ramirez Ruiz and by Carlos Pereyra for two reasons. In the first place because both of them died while I was preparing this book; the second one is that both were very talented. I cannot but feel immensely fortunate to have been Carlos's friend and Ramirez's son. Let this be a tribute to their memory.

SANTIAGO RAMIREZ, May 1995
Department of Mathematics and Department of Philosophy,
Universidad Nacional Autónoma de México,

WALTER BELLER

LOGIC IN MEXICO

The purpose of this work is to show how logic has developed in Mexican society from the middle of the last century until the present day. This process presents two characteristics: (1) works of logic were generally written by authors who, in spite of their national and international prestige, never succeeded in forming a school, but limited themselves to follow the trends outlined by others; and (2) the greater part of the work about logic was written to cover educational needs and, therefore, has been subject to national education policy changes. On the other hand, it should be taken into consideration that the authors we are going to review may be divided into two main groups: those who have used logic from the perspective of some philosophy (i.e. positivism, neo-Kantianism and Marxism) and those who have spread it within the reference framework of formal contemporary logic, whether 'mathematical logic' or 'symbolic logic'. We are going to find that those writers forming the first group have been, for several reasons, opposed to the introduction of mathematical or symbolic logic.

I would like to point out from the beginning that we are not going to cite all of the works of logic published by Mexican authors during the period taken into consideration; instead we are only going to examine some of the books that may be considered the most representative of logic in Mexico. I would also like to indicate that this work is aimed mostly at the global disclosure of the works and their authors rather than a detailed examination. Finally, this study focuses on published works beginning with the social and political movement that is known as the Reformation, the most important social and sociological event after Mexico achieved its independence from Spain.

Although I have limited myself to the study of works written outside the Scholastic influence, it is necessary to remember the case of the renowned piece of literary work created in this land during the colonial period. I refer to *Commentaries on Aristotle's Theory of Logic*, in some publications called *The Mexican Logic*, by Antonio Rubio de Rueda (1548–1615), with no fewer than eighteen editions in the Euro-

S. Ramirez and R. S. Cohen (eds.), Mexican Studies in the History and Philosophy of Science, 1–22.

pean countries during the first half of the 18th century. Rubio, a Jesuit, a Spanish theologian, and a philosopher, was born in Villa de Rueda (Spain), moved to Mexico when he was twenty-eight years old and devoted the last third of his almost twenty-five years in Mexico to writing at least part of his philosophic work. Professor Walter Redmond has carried out an excellent investigation about this distinguished author.[1]

I. LOGIC IN PHILOSOPHIC PERSPECTIVE

1.1. Educational and Political Importance of Logic

Due to the social revolution known as the *Reforma*, led by Benito Juárez, Mexico was transformed at all levels: economic, political and ideological. An educational system was founded then that has continued to the present day. Since 1867, obligatory, lay, primary instruction began, and was paid for by the state. During that same year the National Preparatory School was created, this educational level to be taken prior to college studies, where numerous generations were formed with solid scientific knowledge. The National Preparatory School was born oriented to the positivist philosophy of Auguste Comte, and incorporated logic as a preeminent subject.

Gabino Barreda (1818–1881), who had met Comte while the latter was lecturing in Paris during the years 1848–1851, developed the study plan of the Preparatory School conforming to the scientific system formulated by the French philosopher.[2] Nevertheless, Barreda did not completely copy this system; instead he introduced some changes of a pedagogic order (such as the incorporation of subjects like geography and history, deliberately excluded from the Comtean classification). Studies were arranged following the order that Comte had considered most appropriate, from both the historical and logical points of view, as a series of growing complexities. But what is interesting to point out here is the place and importance the founder of the Preparatory granted to logic. Barreda placed logic after the other sciences. Logic – he wrote – "can only come after the different logical methods and the diverse devices that merit human understanding to reach the evidence that has become familiar to the students, by putting them into practice and then putting them into the scientific studies that they have been through progressively and gradually".[3]

The logic Barreda proposed to introduce was distinguished by two features: on the one hand it should be completely different from the Scholastic educational style, searching for the constant complement of theory and practice; and, on the other hand, it should include 'inductive logic'.[4] For this second aspect, Barreda expressly quotes John Stuart Mill's book called *System of Logic* (1843) which he considered the most adequate for the education of the student of the Preparatory School. Thus, the subject of logic was initiated in modern Mexico following Mill's guidelines. It deals with one conception, as debatable as it may be, that commends observation and experimentation, and rejects whatever interference of metaphysics. Barreda stuck to this conception, since it was compatible with the scientific formation pursued by his curriculum.

The conservative forces opposed, in several different ways, the development of the National Preparatory School. In particular, the subject of logic turned into the center of a parliamentary debate. In effect, in October of 1880, the Minister of Public Education communicated the decision of the then President of the Mexican Republic, Manuel González, to replace Alejandro Bain's book (John Stuart Mill's proselyte), which had been previously chosen by preparatory teachers, with the book of a spiritualist author (Tiberghien). The decision was questioned by the deputies of positivist orientation, one of whom was Justo Sierra (1848–1912), a prominent figure of Mexican culture. "The fight for the adoption of the logic text and the appointment of a teacher for this subject continued with force, mainly during the presidency of Manuel González and the first two years of Porfirio Díaz's second term in office".[5] That is to say, the conflict lasted approximately six years.

Very few times in history has the study of logic incited a parliamentary controversy such as in Mexico!

Logic, then, became a subject of opposition between those who wanted to bring back metaphysics in studies at the preparatory level (and with that to return to a religious education) and those opposed to such a regression. In these circumstances Porfirio Parra's book *Nuevo sistema de lógica inductiva y deductiva* (New System of Inductive and Deductive Logic) was published.

1.2. Porfirio Parra, Man of Science

Parra was born in the city of Chihuahua, capital of the state of Chihuahua, Mexico, on February 26, 1854. When he was eleven years old, he entered the Scientific and Literary Institute of his native city to study mathematics and philosophy. As he proved to be intelligent and keenly alert, the local government awarded him a scholarship so he could continue his studies in Mexico City. When he was fourteen years old (the same age as Aristotle when he joined the Academy) he entered the third course at the National Preparatory School, where he was Gabino Barreda's favorite pupil. While he was still a student in 1871, he obtained, by competitive examination, the professorship of world history and Mexican history. Three years later he entered the National School of Medicine and finished his studies in February 1878. Almost immediately afterwards he joined the 'Hospital Juárez' as a medical doctor. During this period, when Barreda was sent to Germany as Mexican Ambassador, Parra substituted for his teacher in the logic class.

Parra was a prolific writer. As a journalist he founded the journals: *Método* (Method) and *Positivismo* (Positivism). He published many works, among them *Oda a las matemáticas* (Ode to Mathematics) (1887); *Estudios Filosóficos* (Philosophical Studies) (1896); *La colaboración intelectual de Barreda con los trabajos de Juárez* (Barreda's Intellectual Collaboration in the Works of Juárez) (1897); a novel called *Pacotillas* (Junk) (1900); *Discursos y poemas* (Speeches and Poetry) (1908). His book on logic was published in 1903 (the same year that Bertrand Russell published his work *The Principles of Mathematics*).

He was the first principal of the School of Higher Studies (which preceded the current Faculty of Philosophy and Letters of the National Autonomous University of Mexico). He represented Mexico in several international scientific conferences, in Spain, Brussels, Paris and Lisbon. He died in Mexico City on July 5, 1912.

Many authors consider Parra as a positivist thinker, and although this is not untrue, such a characterization is, in fact, insufficient, as under the adjective of 'positivist' one often tends to gather, in the same group, philosophers whose theses diverge, as is the case of Comte, John Stuart Mill and Spencer. As a matter of fact, because of the existence of opposing theses among the positivists, Parra himself opted for the norm of submitting all of them to the 'scientific method' and, as a consequence, to admit only those plans that could be proven and/or incorporated to others already verified. In the words of the Mexican logician:

> We will not affiliate to any specific system, since cultivating the scientific method will be the main object of our tasks; to

> its highly regarded opinion we will submit *all doctrines*, all opinions; those that it sanctions will form an integral part of our unchangeable creed; those that will be incompatible with it will be condemned without pity, regardless of how much sympathy they would have inspired us before test.[6]

Nevertheless, this apparent neutrality hides, as will be shown later on, the position that Parra adopted in favor of John Stuart Mill's thesis of epistemology and logic.

The book *Nuevo sistema de lógica inductiva y deductiva* was inspired by Mill's book and, thus, it is not an original work in the strictest sense, even though it has the merit of having introduced some detailed planning; above all, it is significant for having incorporated to the Spanish languages – in agreement with its specific characteristics – the innovations that Mill and his followers had initiated in the English language.

1.3. Objectives and Structure of the Work

There were two main preoccupations regarding the manual written by Parra: to transmit to the student at preparatory level interest in proving his beliefs, and to show that methods exist that make it possible to advance their knowledge of theory and practice – methods that can be learned and mastered. It deals with a work destined to become popular in teaching the subjects of the theory of knowledge, formal logic, inductive reasoning and methodology.

Nuevo sistema de lógica inductiva y deductiva, a book originally published in two volumes, divided into four basic parts. In the first part, Parra set forth the importance of the study of logic and criticized the speculative versions about logic (among which he includes the Scholastic, as well as writings on logic by Hegel). The following part is devoted to the examination of some questions regarding the theory of knowledge. It also deals with the connection between psychology and logic, as well as the nature of the 'principles of logic'. The third part consists of a detailed study of words in their cognitive meanings. Lastly, in the second volume we find a series of methodological guidelines, as well as the use of inductive inference.

Following a tradition that Mill in turn inherited from the Renaissance logicians, Parra maintained that logic is as much a science as it is an art.

> As science, in establishing the general principles of belief; as art, in formulating the precepts which have to submerge the intelligence to conveniently interpret those principles.[7]

In both cases the purpose of logic is to reach true knowledge, understanding truth as the correspondence between the ideal and the real. Thus, we are faced by a conception of logic which also includes the forms of induction, the theme of the increase of knowledge, and the subject of truth. Consequently, Parra defined logic as the science and art of "acquiring, coordinating and verifying knowledge, aiming at showing the exact congruence between what is ideal and what is real".[8]

On the other hand, Parra admitted that there are truths that require no proof, as their truth is evident to human understanding; such truths sprout others, by generalization, which have to be subdued to proof. The first type receives the name of 'truths of intuition' and they are those that derive either from the testimonies of the senses or, rather, from sensitivity, while the second type is called "truths of inference', which are recognized as truths only after they have passed the test.

1.4. Balance

Porfirio Parra's manual of logic was used in Mexico as a textbook for forty years. As for the system of logic, Parra remained faithful to the Aristotelian syllogistic.

> The doctrine of syllogism – he wrote – was elaborated by Aristotle in all of its essence. Truly amazing is the extreme sagacity and great penetration of this unique thinker, who, from the bottom to the peak, raised his finished monument of human reason. Nothing has been accomplished regarding syllogism during the subsequent centuries; it either has to be accepted as it was created by Aristotle, or cast aside as a curious archeological item. Scholastics have only added details to it, it has withstood criticism all through the 18th century. Among the logicians of the current century, some have scorned it without reason, Mill has rehabilitated it and Hamilton, Boole and other reformers of logic, have not succeeded in altering the essence of this prodigy of human reasoning.[9]

With great acuteness, Alfonso Reyes wrote that Parra did not want to learn about anything new in the subject of logic, which attitude had definitely kept him away from mathematical logic.[10] In effect, Parra was convinced that the syllogistic was a perfect system, to which only details could be added. However, one needs to remember that during the period when Parra's papers were published and studied, no other alternative system of logic was yet commonly known. The works of Frege, Peirce, Peano, Russell or Cantor circulated among specialized groups even in Europe and little or nothing was known of them in Mexico.

Still under influence of Mill's thesis, we find *Nociones de lógica* (Logic Notions), by Samuel García Rodríguez (1862–1957), who, like Parra, studied mathematics and qualified as a medical doctor. In his book García follows the experimental postulates devised by Claude Bernard, considering the principles of deduction simply as by-products and secondary, therefore placing induction as the highest operation in logic. Other books were also published, which aimed at summarizing the fundamental statement of logic in the manner of Mill. One case is Dr. Román Ramírez's *Epítome de lógica, psicología y moral* (Epitome of Logic, Psychology and Morality).

2.1. *The Process of Modernization*

The same year that Whitehead and Russell published the first volume of their *Principia Mathematica*, the Mexican Revolution began. Half way through the decade of the twenties, the revolutionary turbulence began a step towards the consolidation of institutional life. From the thirties on, Mexico experienced a definite transition from rural to urban life. During this time thousands of schools and numerous industrial centers were put into service, as well as hospitals and shopping centers. Light industry and trade also expanded. As for development of education, a progressive increase in student enrollment occurred in both middle and higher education schools; the National Polytechnic Institute was born and different regional technological institutes proliferated, as well as teacher-training schools and universities in several states of the Mexican Republic. Within the Secretary of Public Education, the government's agency in charge of carrying out the politics of national education, the succeeding figures of command were very individual and had very distinct ideologies, Narciso Bassols, Luis Sánchez Pontón, Jaime Torres Bodet and José Angel Ceniceros among others. Regarding philosophy, the intellectual leaders of that time, who had been educated following the thesis of positivist philosophy, took diverse positions: Antonio Caso went for Bergsonian vitalism, Vicente Lombardo Toledano for Marxism, Samuel Ramos for phenomenology and the study of the Mexican personality. The neo-Kantian school, a type of idealism that arose in Germany during the last third of the 19th century, had Francisco Larroyo, who also delved into logic, as its most distinguished representative.

Larroyo knew how to join the theses of the two main schools of neo-Kantianism. As is well known, neo-Kantianism is divided into two different tendencies: the so-called logicist school of Marburg and the so-called axiologic school of Baden. For the first one, every object that can be known is subject to the immanent logical laws of pure reason. Like all neo-Kantians, they reject sensation as an independent factor of knowledge. Sensitivity does not face thought as a strange element; instead, it simply is an unknown which can be compared with the magnitudes to be determined, such as the 'X' in mathematics. It is not something that is given to knowledge, but something 'outlined' within knowledge, which is determined by itself. Reason, for this school, consists in the progressive unfolding of judgment, in such a way that the object of knowledge is the product of this activity, an activity which is conceptual. All of reality remains reduced to an interwoven part of logical relations.

The Baden school turns toward the study of culture, whose foundations are not laws of logic, but 'axiologic', based on values. It establishes that if judgment of reality has pretensions to truth and objectivity, this is due to the presence of transcendental values, which involve the fulfillment of a duty to be so. Thus, the scientist who with his activity tries to establish true statements about the reality that he studies, is fulfilling a value, the urgent need to arrive at the truth.

2.2. *Francisco Larroyo, Philosopher and Neo-Kantian*

Larroyo was born in Jerez, Zacatecas, México, in 1912. His full name was Francisco Luna Arroyo. He entered the National School for Higher Studies (nowadays the Faculty of Philosophy and Letters of the UNAM). In 1929, he obtained a scholarship to study in Germany, where he attended lectures by Rickert, Husserl and Liebert, among others. Three years later he returned to Mexico and obtained a master's degree and a doctorate of philosophy at UNAM, as well as a master's degree in the science of education. In 1945 he formed part of the faculty that founded the Normal Superior School, a teacher-training school where they trained teachers-to-be and gave them the skills to impart basic education. He later occupied other positions within the Secretary of Public Education and in 1958 was named Principal of the Faculty of Philosophy and Letters. He died in 1981.

With the objective of spreading the ideas of neo-Kantianism, Larroyo founded *La Gaceta Filosófica* (The Philosophical Journal). His own bibliography was extensive: he published more than twenty-five books about philosophy, pedagogy, history of ideas, etc., besides having translated several important philosophic works: *La filosofía de los valores* (The Philosophy of Values) (1936); *Los principos de la ética social* (Principles of Social Ethics) (1936); *La lógica de las ciencias* (The Logic of the Sciences) (1938) (in collaboration with Miguel Angel Cevallos); *Historia general de la pedagogía* (General History of Pedagogy) (1946); *La antropología concreta* (Concrete Anthropology) (1958); *Sistema de la estética* (System of Aesthetics) (1966); *Filosofía de las matemáticas* (Philosophy of Mathematics) (1976).

In his book *La filosofía de los valores* Larroyo announced that he was going to write a text "as an introduction to current logic, a book conceived with the latest contributed ideas". He said that he was trying to compare the principles of traditional logic and, on the other hand, defend a 'dialectic' position. The following year Larroyo published *La lógica de las ciencias* (1937), which was a first approach dealing only with problems related to the founding of logic. By 1939 the work acquired a more definite structure.

By the fourth edition of the work, Larroyo introduced new elements; he incorporated a chapter about the cause and principle of Heisenberg's indeterminism. The tenth edition (1959) presents new chapters about 'the logic of mathematics', 'the logic of physics', etc. Not until the sixteenth edition (1969) does it incorporate some aspects of 'symbolic logic'. It should be remembered that from its first edition, this work was intended to serve as a manual for mid-level teaching.

Larroyo's manual little by little replaced Porfirio Parra's logic book.

2.3. *Objectives and Structure of the Work*

The main objective of Larroyo's book is to postulate the inseparable relationship between logic and the different particular sciences, pointing out that logic advances only when the latest results of scientific investigation are incorporated. *La lógica de las ciencias* claims to be a (philosophical) theory of scientific work. For Larroyo, logic is the science that studies the laws to which scientific thought is subjected. Logic

> explores the forms of thinking; tells of the organization and structure of scientific thought; speaks of the nature of thought by way of which sciences are elaborated and established; accounts for and explores the methods followed by the investigators; of the essence of scientific hypothesis, definitions, classifications and the outline of problems, etc. Thus, it is understood that logic is basically a *theory of the sciences*: in short, a discipline which describes and explains the common forms of scientific thought, that is, the contextual logic of the diverse sciences.[11]

But being that in such characterization logic is not detained in any particular science and instead looks for general regularities, it is an eminently philosophical discipline. And being that philosophy has culture and cultural values as its object of study, logic explores a basic value: the truth.

The work is divided into two main parts. The part that Larroyo calls "*Apofántica*" deals with the examination of traditional logical operations: the concept, the judgment, and the inference (deductive, inductive and analogical). The most important part here is the explanation of judgment as "an operation of thought thanks to which a matter of knowledge is determined from a point of view that fixes the sense of a question".[12]

That is to say, any judgment is the answer to a question which the subject asks himself. Since questions play a fundamental role in that activity, Larroyo includes in his work a part regarding *aporetic* or the 'logic of interrogation'. And given that that science, as well as any other 'cultural territory', fulfills a norm, Larroyo also includes a part about the 'logic of the norm'. Likewise, he examines some 'philosophic methods', such as: neo-Kantianism, dialectical materialism, phenomenology, the neopositivist semiotic, and structuralism.

In the other main part of *La lógica de las ciencias*, the author deals with the *theoretic*, in which he tries to account for the combination of the essential points of any theory, highlighting its inseparable elements. Here he repeats that science has two unmistakable themes: science involves a *system* because each one of its 'truths' is subordinate to a determined order in such a way that all its propositions form a unity; science establishes *theories* because it gives the necessary proof of knowledge, demonstrating that some truths are derived from others.[13]

Larroyo then proceeds to present diverse classifications of science and ends by showing how, according to him, logic is present in the different sciences: mathematics, physics, chemistry, biology, as well as sociology, history, political economy, and even geography, including the sciences of law and ethics.

2.4. *Balance*

Larroyo's book contributed to a certain advancement in the knowledge of logic, above all for its criticism of the traditional explanation of judgment, backed, as Larroyo shows, by grammatical elements. However, this author does not succeed in proposing a coherent and alternative logical system. Larroyo was equipped with the outlines of various logicians such as Edmund Husserl, Hermann Cohen, Paul Natorp and Alexander Pfander, among others; but precisely for that reason he did not succeed in building a unitary and systematic conception (characters which Larroyo himself demanded for an investigation with scientific pretensions). Moreover, the structure of the book, carried out on the basis of further incorporating 'last-minute' information results in a disorganized and unconnected ensemble.

On the other hand, Larroyo tried to show the limits of formal logic, indicating, for example, that the logical principles of identity and of non-contradiction are insufficient for the advancement of knowledge, and he claims to surpass them by using the principles of neo-Kantian philosophy, with which he arrives at a series of senseless ideas which are the consequence of a confusion of levels in the analysis of knowledge. Moreover, Larroyo's book includes a series of prejudices and absurdities about mathematical and/or symbolic logic; for example, he considers that a logic of tautology is perfectly useless for scientific knowledge. Reckless and unfounded allegations, confusing and misleading, are what we find in this book that, to this date, has had more than twenty editions. Larroyo believed that symbolic logic is no more than a formal dressing of the traditional syllogistic inference. Prejudiced opinion prevented him from understanding the innovations that contemporary logic brought with it.

3.1. The Search for a Scientific Philosophy

During the decades of the fifties and sixties Mexico was involved in contradictory economic processes. In 1960 President Adolfo López Mateos nationalized the electrical power industry, with which the process that started with the expropriation of the oil industry continued (this had been achieved by President Lázaro Cárdenas in 1938). These measures apparently enabled the Mexican state to handle its basic-economy activities, and were known as 'the economic management of the state'. On the other hand, industrial production increased, mainly in consumption goods. Since then the overpowering presence of foreign industries, whose factories operated in Mexico as branches of the great transnational enterprises, mainly North American, was already noticeable.

In another direction, a definite and irreversible consolidation of urban life took place, although its most active dynamism was in Mexico City, monopolizing the greater part of the cultural and scientific activities of the country. The philosophical preoccupations were gradually displaced from existentialism (in all its variants) to the philosophy of science, and within the latter, two mainstreams have existed: on the one hand, Marxist philosophy, and on the other, the philosophy derived from the Vienna Circle.

Even though Marxism was present in university life in the voice and thought of Vicente Lombardo Toledano, it acquired more strength during the sixties. Two works were written with the purpose of making original contributions about Marxism. The *Introduccíon a la lógica dialectica* (1956), by Eli De Gortari, and *Filosofía de la praxis* (1967), by Adolfo Sánchez Vázquez, a Spaniard who had taken root in Mexico.

Before examining the work of Eli De Gortari, it is useful to glance quickly at the more general outlines of dialectical logic. We can say that the nucleus of the materialistic dialectic (at least in its most orthodox version) is constituted by three 'laws' which Engels arrived at in his well-known polemic with Dühring: the identity and struggle of opposites, the passing of quantitative into qualitative, and the law of the negation of negation. There is no standardized uniform formula regarding these laws, nor is there a single nomenclature. Besides, they are not to be represented formally. We will stop with the first of these laws in order to have arguments for the balance of this work.

The content of the law of identity and struggle of opposites is the following: everything and every process contains opposite elements that mutually exclude one another and for this reason they are in reciprocal contradiction. These opposites constitute, at the same time, a polarity so that they presuppose and reciprocally influence one another, thus forming a dialectical unit. This unit, and this struggle of some internal opposites, is the propelling force of transformation and advances. The resolution of the contradiction is parallel to the progress of a thing at a new evolutionary level. In dialectical terminology, this conflict of opposites is also called 'objective contradiction'.

On the other hand, dialectical logic as a logic based on dialectical principles, has been the other object of violent controversies, above all when one wants to oppose it with formal logic. For several years, Marxist philosophers debated the question. At first they admitted the existence of a formal logic, as well as that of a dialectical logic, but not as two separate logics. According to this, formal logic deals with the simple static phenomenon in an abstract and fixed way. On the contrary, dialectical logic deals with movement and change, and it also deals with the complexities of a reality that is in evolution. For this reason the conception of the dialectic as a superior form of thought was reestablished.

In a second conception by other Marxist philosophers, it is affirmed that two logics do not exist – one formal and the other dialectic – but only one: formal logic, which deals with correct reasoning, while dialectical logic is no more than dialectic and coincides with the theory of knowledge, as Lenin stated in his *Philosophical Notebooks*.

3.2. Eli De Gortari, Dialectical Philosophy

Born in Mexico City in 1918, De Gortari studied engineering, mathematics and philosophy, obtained a degree as a municipal and sanitary engineer, and a master's degree and a doctorate in philosophy at UNAM. He was rector of the Michoacán University of San Nicolás de Hidalgo (1961–1963). He was an 'official investigator' at the Institute of Philosophic Investigations of UNAM and a professor (since 1948) at the Faculty of Philosophy and Letters, also of UNAM. He was the Seminar Director of Scientific and Philosophic Problems, which gave rise to numerous publications on philosophy and science, besides having put him in contact with scientists and philosophers, both local and foreign. In fact, this important seminar stopped functioning in 1968, when De Gortari was jailed by the same government that massacred young students on October 2, 1968. He was released two years later.

The bibliography of De Gortari is abundant; following is a selection: *La ciencia de la lógica* (The Science of Logic) (1950; 2nd Ed. 1979); *Introducción a la lógica dialéctica* (Introduction to Dialectical Logic) (1956, 4th Ed. corrected and augmented, 1972) – this work was translated into Russian in 1959 – *La ciencia en la historia de México* (Science in the History of Mexico) (2nd Ed. 1980); *El método dialéctico* (The Dialectical Method) (1970); *Lógica general* (General Logic) (5th Ed. 1972); *Iniciación a la lógica* (Introduction to Logic) (2nd Ed. 1972); *El método de las ciencias* (The Method of the Sciences) (1970); *Fundamentos de la lógica* (Fundamentals of Logic) (1982); and other scientific books, plus essays about the history of science in general and in Mexico, as well as his *Dialéctica de la física* (Dialectic of Physics) (1964).

Based on Boole's algebra, De Gortari published in 1955 the text 'The deductive phase of the materialist dialectical method', which appeared in a journal called *Dianoia* (Number 1). In this work he submitted sixteen proposed forms to show his related graphical representations and offered some examples applicable to different sciences. De Gortari followed the same outline as Larroyo: Logic must be linked to the latest scientific achievements. In 1956 he published his *Introducción a la lógica dialéctica*, which in its first edition only incorporated the text already quoted and some considerations regarding the fundamentals of logic. For its fourth edition, 'theory of the transductive inference' was added to the book, together with other sections devoted to the graphical expressions of the forms of judgment, and about the logical estimate of inferences. The books *Lógica general* and *Iniciación a la lógica* are no more than corrected versions, enlarged and more didactic, of the *Introducción a la lógica dialéctica*.

3.3. *Objectives and Structure of the Work*

This text is composed of twelve chapters in which De Gortari gives a detailed account of the mastery of philosophy and of science, examines the dialectical laws, and proposes a materialist dialectical interpretation of the concept, judgment and inference (inductive, deductive and transductive), concluding with the exposition of some elements of scientific method. De Gortari defines science by saying that it is "the objective and rational explanation of the universe".[14]

Meanwhile philosophy, which for him was also scientific because it offers objective and rational explanations, deals with knowledge in general, inasmuch as it fulfills the general elements of acquired scientific knowledge.

The interpretation that De Gortari sustains about logic is ample:

> Logic – he wrote – is the philosophical discipline that deals with the formulation of the methods of scientific investigation and, for that purpose, it is up to it to analyze the process of thought required to discover the forms that the elements of thought adopt, the functions that link them, the methods employed in the investigation and the laws of theoretic and experimental knowledge.[15]

For De Gortari, logic can be verified, modified or refuted when it is applied to the process of knowledge, and therefore logic has the objective of studying scientific knowledge, and this, in turn, is its source of resistance.

As for concept and judgment, De Gortari gives them a similar treatment to that of Hegel in his *Wissenschaft der Logic* (*Science of Logic*); that is to say, the *concept* is presented as a moment inside the 'dialectic' of knowledge, overcome in the superior form of *judgment* which, in turn, is product of the connection of two diverse terms. "Consequently, the same judgment contains the essence of a contradiction, since it relatively identifies one term with another diverse term".[16]

Regarding deductive inference, De Gortari stays at the syllogistic level, but with a treatment that is supported in the work of George Boole, *The Mathematical Analysis of Logic, Being an Essay Towards a Calculus of Deductive Reasoning* (1847). In this manner, to the nineteen valid ways of traditional (Aristotelian) syllogism, he could add one hundred and sixty-five more, to make a total of one hundred and eighty-four valid ways in the category of inference.

In his 'transductive' inference, De Gortari established a system of inferences which does not go from the general to the particular – as in the case of deduction – nor from the singular to the general – as in the case of induction – but instead his premise and conclusion remain at the same general level. In reality, it deals with a combination of inferences which fulfill the laws of transitivity and whose premises involve relations of equality or of inequality, or of properties similar to one or another.

The Mexican philosopher understands that formal logic is different from dialectical logic. The former studies the ways of correct thought, for which it makes abstraction of the development and transformations of the processes of reality, keeping the relatively stable aspects of such processes. Formal logic shows us how to think correctly, but it only allows us to arrive at possible results, to formally correct hypotheses that ought to be subjected to experimentation so that, besides their formal validity, they are transformed into real ones. Dialectical logic, on the contrary, studies scientific knowledge in its integrity, in its evolutionary development and in the unfolding of the thought that it reflects. The scope of dialectical logic is much greater, being that it consists of, according to De Gortari, "the laws of thought, those of knowledge and those of objective existence, as well as the interrelations between one and another, resulting that formal logic is a particular case of dialectical logic".[17]

3.4. Balance

The logical work of De Gortari signified a considerable advance in regard to that of Larroyo, for two fundamental reasons:

1. Larroyo's was kept within the limits of one fixed interpretation, so that it keeps its coherence and unity;
2. The work offers some novelties in the exposition of certain contemporary logical themes (such as Boolean algebra). On the other hand, De Gortari frequently reiterates that philosophical reflection should yield to the dictates and corroborations of science, thereby claiming to get rid of excessive speculation in the field of philosophy.

Of course, in another aspect, the work of this Mexican philosopher contributed (during the fifties and sixties) to the argument concerning dialectical logic.

De Gortari's logical work is polemical and could not be otherwise, because it deals with a product of controversial philosophy. Several aspects of this could be discussed, but we will concentrate on only one: the dissertation that formal logic is a particular instance of dialectical logic.

In the first place, the command of 'dialectical logic' is not clear or differentiated; the principles that give life to it are ambiguous and vague. If by 'dialectical logic' we understand a theory concerning reality and its laws, the result is that we find ourselves in the field traditionally called ontology. Now, for one discipline to embrace another in order for one law to become the particular case of another, a necessary condition would be required for both disciplines to share the same references.

Strictly speaking, formal logic is related to reality, but not directly, immediately or evidently; instead it is linked to it through language. The laws of formal logic have been abstracted from language. Therefore, formal logic has language as its 'raw material' and its subject of study is the *relation of logical consequences*. This relation is established exclusively between statements. A statement is, as we know, a linguistic expression of which we can say whether it is true or false, but not both things at the same time. A statement is true when it really describes a fact and is false in the reverse case. It does not fall to formal logic to decide about the truth or the falsity of statements.

To consider that formal logic is a part of a theory of reality (ontology) as De Gortari claims, implies lack of knowledge of the current situation of logical theories. If formal logic constitutes a frame or structure that serves both mathematics and physics or economics, logic itself does not try to go further than its strict formal limits. That is to say, formal logic does not say anything about reality. For that reason, it does not share the same supposed references of dialectical logic.

II. LOGIC IN THE PERSPECTIVE OF SYMBOLIC LOGIC

Symbolic or mathematical logic appeared late in Mexico. Both Mexican mathematicians as well as philosophers ignored this science for many years. This is not surprising if one takes into account that the same situation took place in other places. E.T. Bell, in his work *The Development of Mathematics*, tells how, as late as 1939, a distinguished liberal North American philosopher proclaimed joyfully that mathematical logic had not contributed anything new, neither to mathematics, nor even to the theory of knowledge, for which reason such a 'wise individual' claimed having saved himself the 'bother' of learning such 'unnecessary symbolism'.

One of the first known books about the subject in Mexico was the manual written both by José Ferrater Mora and Hugues Leblanc: *Lógica matemática* (1st edition 1955). Even though its authors are not of Mexican nationality (the first is Spanish and the second Canadian) it is a publication published in Mexico that has been widely talked about, not only among a good number of Mexicans interested in this discipline, but also amongst other Latin Americans and Spaniards equally interested in it. "The work is useful and clear and complies with its informative mission, although the beginner reader has difficulty in appreciating throughout the book the unity of logic in its different sections".[18]

Chronologically speaking, prior to this last work there were texts of Eduardo García Maynes concerning the logic of law. In 1951, his *Introducción a la lógica jurídica* (Introduction to Legal Logic) appeared, the same year that Klug published in Berlin his *Legal Logic* and Georg Henrik von Wright his *Essay about Modal Logic* and two years before Kalinowski published his *Theory of Normative Propositions*. García Maynes tried to apply the resources of mathematical logic to the field of law and ethics, both to problems relating to the Aristotelian idea of practical syllogism and to the question of the validity of judicial reasoning, as well as with the deontic manner that refers to the structure of orderly law.[19]

He later published: *Lógica del juicio jurídico* (Logic of Legal Judgment) (1955); *Lógica del concepto jurídico* (Legal Concept of Logic) (1959); *Ensayos filosófico-jurídicos* (Philosophical-Legal Essays) (1959, 2nd Ed. corrected and augmented, 1984); *Lógica del raciocinio jurídico* (Logic of Legal Reasoning) (1964); among other publications. The work of García Maynes has awakened interest among those devoted to legal logic or to normative logic.

In translated material we find a variety ranging from manuals to classics in the logical literature. Francisco Larroyo had translated *Theory of Sets* by Kurt Grelling (together with Alfonso Juárez). Eli De Gortari translated in 1952 *A Preface to Logic* by Morris R. Cohen. Then, Robert S. Hartman and José Luis González translated, in 1965, L. Susan Stebbing's *A Modern Introduction to Logic*. In 1966 Professor Gabriel Aguirre Carrasco translated Patrick Suppe's book, *Introduction to Logic*. In 1972, Hugo Padilla translated several texts by Gottlob Frege, all of which were embodied in a book also edited by UNAM with the following title in Spanish: *Conceptografía. Los fundamentos de la aritmética y otros estudios filosóficos* (Conceptography: Fundamentals of arithmetic and other philosophical studies).

Some manuals devoted to teaching began to be published. The mathematician Gonzalo Zubieta wrote a book entitled: *Manuel de lógica para estudiantes de matemáticas* (Manual of Logic for Mathematics Students) (1968), in which he tried to make accessible for mathematics students the use of efficient methods of reasoning. Javier Salazar Resines, an engineer, published in 1970 a two-volume work entitled: *Introducción a la lógica deductiva y teoría de los conjuntos* (Introduction to Deductive Logic and the Theory of Sets), which is what students of the bachelor's degree level should begin with in those subjects which required a 'programmed text' for exercise work. Gonzalo Zubieta published in 1973 a short work for students of bachelor's degree level entitled *Lógica elemental.* Francisco Zubieta published a manual entitled *Lógica matemática elemental* (1977). Eli De Gortari wrote in 1983 his *Elementos de lógica matemática*, which claims to be a manual for those being initiated in the subject.

Lastly, it must be recognized that since the sixties, diverse works regarding the philosophy of logic, metalogic, and epistemology of logic, by Mexican investigators, most of whom had studied these subjects abroad, have been written. UNAM's Institute of Philosophic Investigations has a section devoted to this. The UNAM has a Master's degree program in logic. In sum, the work developed at this level now has little relation to the level outlines of the overall group of texts that were analyzed in the first part of our work herewith.

POSTSCRIPT ON ELI DE GORTARI

On the 30th July, 1991, the Mexican philosopher Eli De Gortari died in Mexico City at the age of seventy-three after a distressing illness. He had been one of his country's most outstanding intellectual figures. In this paper we have examined the chief aspects of his work as a logician. Yet, I must now refer to other aspects of his life, as a well-deserved tribute to his educational, scientific and cultural work.

Eli De Gortari distinguished himself as a nonconformist thinker, as much socially as philosophically and scientifically. Socially, because he always maintained a left-wing position, repudiating the unjust economic order which prevails in Mexico and the scientific and technological backwardness imposed on his country. His book *La ciencia en la historia de México* (Science in the History of Mexico) was written to show the possibilities and the socio-political limitations to which Mexican science and technology have been subjected. In like manner, De Gortari repeatedly defended students at various universities who had been under attack by reactionary groups. Because he supported the student movement of 1968, he was jailed as a political prisoner by the administration of President Diaz Ordaz.

As a philosopher, he was a tenacious critic as much of the idealist as of the positivist trends which, from the thirties onwards, sought to dominate the course of the philosophy of science and of scientific activity itself. The project of carrying forward 'dialectical logic' (regardless of the opinion we may have of it today) and a Marxist view of science should be understood as a critical alternative to those trends. This point is important because the penetration in Latin America of positivist ideas was extraordinarily persistent. Under these conditions, many philosophers, researchers and thinkers who were not convinced of the relevance of those ideas found in the philosophical works of Eli De Gortari a different and promising course.

On the other hand, Eli De Gortari's powerful personality manifests itself in his outstanding academic activities. He was a devoted and committed university professor and taught at the most important institutions of higher learning. At the National Autonomous University of Mexico (UNAM), he taught at the Faculties of Arts, Engineering, Economics, Chemistry, Medicine, Science, Psychology and Political and Social Sciences. At the National Polytechnic Institute, he was professor at the National School of Biological Sciences and at the Superior School of Medicine. He also held teaching posts at other universities in Mexico and abroad. Thus, he was justifiably able to write in the preface to his book *El Método de las Ciencias (Nociones Preliminares)* (The Method of the Sciences – Preliminary Notions):

> (the author) has gathered the very rich experience of the many courses and seminars on methodology that he has taught at the bachelor's, master's, doctorate and post-doctorate levels during the past 26 years at a great number of schools, faculties and centres in the most varied fields: philosophy, social sciences, biology, chemistry, medicine, engineering, economics, archaeology, history, anthropology, administration, ethnology, physics and mathematics.

He was also a tireless promoter in Mexico of contemporary philosophic and scientific culture. Together with the philosopher Samual Ramos and the internationally famous astronomer Guillermo Haro, he organized the 'Seminar on Scientific and Philosophical Problems', whose activities started on the 21st of February, 1955, and continued (with involuntary pauses) practically until the Mexican philosopher and logician's death. The principal objective sought by the Seminar was to achieve collaboration between researchers of almost all fields of learning for a common cause: the elucidation of problems of general interests to scientists and philosophers, with the determination of carrying out a critical analysis of those problems and an accurate examination of their possible solutions. The 'Seminar on Scientific and Philosophical Problems' performed a very important task in the area of publications, known and appreciated in many countries. The work of outstanding researchers such as Max Planck, Philipp Frank, Nicolai Lobachevksi, V. Gordon Childe, among many others, was published along with that of the majority of the most important Mexican researchers, either through Spanish translations of fundamental scientific works or through the publication of original research. In this manner, an unprecedented effort in the area of scientific dissemination was initiated in Spanish-speaking countries (a dissemination which, at present, proceeds in a consistent manner only in Spain).

De Gortari also created and promoted a programme for a Master's Degree in the Methodology of Science, from 1975 onwards. This study plan is now operational, with some modifications to the original project, in a section of the National Polytechnic Institute (IPN). It would be well to point out that it is the first Master's in that field in Mexico, and one of very few in Latin America.

With the death of Eli De Gortari, a cycle of the philosophy of science in Mexico has perhaps come to a close; a cycle characterized as much by the passionate discussion of ideas as by a full commitment on behalf of the thinkers to the noblest causes of social well-being and justice. The philosophical task carried out by De Gortari must now be submitted to the judgment of time. The Marxism he defended has a more and more uncertain destiny in a world dominated by 'free enterprise'. Some of his academic contributions will remain, and so will, of course, the example he set of dignity and intellectual integrity.

NOTES

[1] Cf. Walter Redmond, *Bibliography of the Philosophy in the Iberian Colonies of America*, The Hague: Martinus Nijhoff, 1972. 'Logic and Science in *Rubio's Mexican Logic*', in *Quipu* (Latin American Magazine of History of Science and Technology) January–April, 1984. pp. 55–82.

[2] Cf. A. Comte, *Cours de philosophie positive*. 2 Vols. Paris: Hermann, 1975.

[3] G. Barreda, *Estudios*, UNAM, Mexico, 1978. p. 25.
[4] *Vid. ibid.*, pp. 27–30, see, as well, W. Beller, B. Méndez, S. Ramírez, *El positivismo mexicano*, México, UAM, 1985.
[5] D. Cossío Villegas, *et al.*, *Historia Moderna de México. El Porfiriato*, México: Hermes, 1973. p. 612.
[6] P. Parra, *Anales de la Sociedad metodófila G. Barreda*, México, 1978.
[7] P. Parra, *New System of Inductive and Deductive Logic*. I. 'Economic Typography'. 1903. p. 24.
[8] *Ibid.*, p. 236.
[9] *Ibid.*, p. 236.
[10] Cf. Alfonso Reyes' 'Passado Inmediato' in *Obras Completas*, Vol. XII. México. Reprint 1980. p. 190.
[11] F. Larroyo, *La lógica de las ciencias*, México: Editorial Porrúa. 18 edition. 1973. pp. 34–35.
[12] *Idem*, p. 126.
[13] *Ibid.*, pp. 327–329 and ss.
[14] E. De Gortari, *Introducción a la lógica dialéctica*, México: Fondo de Cultura Econónomica. 4th Ed. 1972. p. 13.
[15] *Ibid.*, pp. 21–22.
[16] *Ibid.*, p. 139.
[17] *Ibid.*, p. 28.
[18] Alfredo Deaño and J an del Val, 'El desarrollo de la lógica en España durante los últimos años', in A. Deaño, *El resto no es silencio*, Madrid: Taurus. p. 323.
[19] F. Salmerón in *Las Humanidades en México, (1950–1975)*, México: UNAM, 1978. pp. 240–244.

JOSE RUIZ DE ESPARZA

CONTRIBUTIONS OF SANTIAGO RAMÍREZ TO THE HISTORY OF THE *COLEGIO DE MINERÍA*

The purpose of this paper is to enhance the works of Santiago Ramírez, a man of whom little is remembered and yet it is through his own carefully researched volume that we have a great deal of historical information about the institution for which he cared most: the *Colegio de Minería* (College of Mines). It is worthwhile for us to look upon some details about his life in the College.

He recorded these personal data in his book *Datos para la historia del Colegio de Minería* a facsimile of which has recently been published by the Engineering School Alumni Society.[1]

Santiago Ramírez was a distinguished pupil. Born in Mexico City in 1841, he should have enrolled in 1857 since he appears as a second year student in the course of mathematics in 1858. As an intern he soon started to share responsibilities as chief in his section. In 1859, during the annual public awards celebrations, before various distinguished figures including the President of the Republic, he read a poem. When prizes were awarded in 1860, he again appeared to read a poem. In 1864 he received his degree as a mining engineer. That same year he was already an officer in the school, namely dean of studies and substitute professor. In this capacity he signed the announcement for the opening of courses.

It was also in that year that he became professor of "perhaps the most important subject in the school, Mineralogy". At the end of the courses, prizes were conferred at a single ceremony for the schools of mining, medicine and agriculture. Among those present were the Emperor and representatives of scientific and literary societies as well as members of the diplomatic corps. Santiago Ramírez delivered the keynote address in a ceremony that the press judged "one of the most brilliant of our times".

Two years later, still a dean, he administered the affairs of the Practical School of Mines (*Escuela Práctica de Minas*) and, again, delivered the commencement address. In 1867 hard times forced most of the students to leave the school by June 20, and Santiago Ramírez lectured for the

S. Ramirez and R. S. Cohen (eds.), Mexican Studies in the History and Philosophy of Science, 23–28.

last time in Mineralogy only days before the liberal army occupied the College. A month later, the building became a military prison for the supporters of the Empire. At this point Ramírez ceased all his direct participation in the *Colegio de Minería*.

Most likely, it was during those years that he organized and studied the Archives of the College. Ten years later, and for the following thirty, he published several biographical studies of those whom he considered to be the most significant personalities in the formation of the College, and also his most important work, the *Datos para la historia del Colegio de Minería*, published in 1890.[2]

The lives of these personalities were of special importance for Santiago Ramírez. The first to be published, in a series that was to be a 'Gallery of Mexican Miners', was that of Joaquín Velázquez de León in 1885; three years later he published the life of Joaquín Velázquez Cárdenas y León, in 1889, Manuel Ruíz de Tejada, two years later the biography of Manuel Andrés del Río, in 1900 that of José Joaquín Arriaga, and the following year a funeral eulogy of Mariano de la Bárcena. In 1902, he published a biography of Ignacio Alcocer, one of Miguel Velázquez de León, and in 1903 that of José María Alcocer.

In each case he undertook exhaustive study, and overcame serious difficulties – such as obtaining certified documents from their native places, or lists of press articles that mentioned his subject. In some cases he gives us references as to where to find complete biographies or he appends documents to prove the quality of the work of his subject. He never fails to establish the connections they had with the College.

Santiago Ramírez had a long life that spanned the second half of the 19th century until the third decade of the 20th; on January 17, 1921 he died in Atzcapotzalco.

While the purpose of this paper is Santiago Ramírez's interest in the College and its history, we cannot neglect how his perspective embodied all the problems of Mexican mining, as can be witnessed in his 1884 work, *Noticia histórica de la riqueza minera de México y su actual estado de explotación*.[3] This includes a historical study of Mexican mining development since the Conquest, based on a broad bibliographical analysis and personal reconnaissance of the Mexican territory.

The focus of his *Datos para la historia del Colegio de Minería* is extremely interesting. One of his main objectives is to stress the importance of the College for the country, in terms both of the works it promoted and of the talent it produced. Ramírez's argument shows that before the establishment of the *Real Seminario de Minas*, Mexican mathematicians were inadequate to meet the needs of national development. This situation would improve due to the influence of the Seminar; he does not forget the first rate scientific collection contained in the Seminar's library, its modern instruments, and the wealth of its collections of classified minerals.

In terms of the human resources produced by the College, he notes that among the alumni there were several justly famous professors, geographers, chemists, miners, metallurgists, writers, and men who were prominent in science, politics, diplomacy and parliament. Above all, the College produced the first Mexican engineers to develop practical solutions to all kind of problems, solutions that were to have a profound impact in the development on Mexican industry.

His vision about the significance of the College and its social influence allowed Ramírez to include information about national life and about the difficulties that Mexico necessarily had to face during the first years of its independent existence, difficulties that were reflected within the life of the College. It is interesting to note that the speed for the creation of a special archive on mining was already evident for Santiago Ramírez. He dreamed of an organization that, among other tasks, would gather all materials that the government kept on mining. Beyond that, he judged it imperative to publish a periodical on the subject.

Santiago Ramírez held different posts in the College which gave him valuable insights into its functions, possibilities and, of course, its difficulties. No one could have known the management of the College better. He was an intern, chief of section, assistant professor and dean of studies (a post immediately under the director), member of its governing board and Professor of Mineralogy. As he says, these gave him a deep feeling of communion and love toward the College.

Perhaps this was what motivated him to write first the history of the *Real Seminario*, and then of the College. The task was literally impossible since the archives and libraries to be visited were directly opposed to the working visits he had to carry out through the country. Thus, he deemed the publication of his work to be useful eventually for somebody who would be interested in the subject.

This happy event allows us to reconstruct the academic life of the College. In the pages of his work, we find practically every indication needed for knowledge of the various programs, their span, their reforms and even the suggestions that never became real.

It was also a part of his goal to update the relevant bibliography for the purpose of such a history. Some examples:

On February 16th, 1859, an alumnus started publication in the daily journal *Diario de Avisos* of a commentary about the work of Juan Bowring on the uses of chemistry and electricity in the processing of silver minerals.

On November 3rd of the same year, Francisco Díaz Covarrubias printed his calculations of the geographical position of Mexico City.

On January 1, 1860, Joaguín Velázquez de León started the publication, in the printing shop of Andrade and Escalante, of the *Anales Mexicanos*. This periodical would become a thick volume with the works of students of the College.

On March 1, 1860, the *Anuario del Colegio* was published. It publication had been interrupted in 1849. Among other things, it contained papers on the law that reformed the College on December 20, 1858, the study program for the Practical School for 1860 and the speech of a professor of mechanics dealing with the conical pendulum or centrifugal force regulator.

On April 17, 1861, Joaquín Velázquez de León and Pascual Arenas started publishing the *Anales de la minería mexicana*. Ramírez appends the press commentaries that appeared in *El Heraldo*, *La Unión Federal*, *El Movimiento*, *El Monitor Republicano*, *El Constitucional*, *El Despertador del Pueblo*, *El Hombre*, *La Restauración*, *La Convención*, *El Indio*, *La Crisis*, *La Epoca* and many others.

Ramírez was also interested in the laws and decrees concerning the College even if the relation was not direct. For example, when on March 17, 1857, the Ministry of Industry demanded that all the professors swear to support the Constitution and the incidents occurring when some of them refused; or when President Comonfort decreed on March 15, 1857, the use of the metric system from January 1, 1862 and the approval of the study program for careers in architecture and mastery of engineering.

There are, of course, other laws that reveal the life of the College even if they were never enacted. On December 20, 1858, a decree was issued to regulate the specialities of mining engineer, topographer, geographer and metal separator. The law of Public Instruction of 1861, the decree of the State Department that arranged the functions of the different Ministers assigning the College to the Ministry of Industry. Or, further, the decree reorganizing the College and the Practical School. These are only some examples among the immense collection of legal acts decreed by the many Mexican governments that Ramírez collects.

Another point of his interest was the relations between the students and the authorities of the College. Thus, during the years in which he was a student himself, he reports certain incidents: on June 1, 1858, three students were expelled for non-religious attitudes; on July 17, six students demonstrated politically "with evident disorder" and did not accept the reprimand passed upon them, and the director prohibited their entry into the school; but they ran over the watchman with the support of the majority of the other students. The minister then ordered the troops to arrest them, and the students, except for fourteen, took over the street, thus forcing the minister to close the school. Two weeks later everything went back to normal after the dismissal of the leaders. On August 16, 1859, a group of external students was expelled from the French course and were asked to take the subject elsewhere.

But not everything consisted of difficulties between students and authorities: on March 31, 1863, professors and students signed a letter against the French intervention. The letter was sent to the minister who fired all the signees.

Our aim has been to show the extremely rigorous documentary selection bequeathed by Santiago Ramírez. It is also through him that we know that he studied the archives of the Mexican Society of Geography and Statistics, the *Sociedad Científica Antonio Alzate* and some of the materials that now constitutes the General National Archive. He also searched in private libraries, such as the one belonging to Luis García Pimentel.

Briefly, then, Santiago Ramírez initiated the task of ordering the prime materials for the history of the *Colegio de Minería*. This labour took him beyond the limits of Mexican engineering. His major accomplishment was to draw our attention toward the subject, and to preserve a set of data without which it would be impossible to estimate the ideas floating in the air during the first decades of existence of the College.

The copious information provides the details that allow us to recover the names of important actors, study the programs, and the legal actions of the government to reorganize or to perfect the College, actions of students, professors, and authorities, convergent at times, divergent at others.

In our times, no one can study the history of Mexican engineering without knowledge of the work of Santiago Ramírez. In the near future, we might come to know other efforts of this illustrious Mexican, words which have never been published but are kept in the historical archives of the school.

I end by quoting the last paragraph of *Datos para la historia del Colegio de Minería*:

> I have but one thing left, to vindicate once more the need for writing the history of the *Colegio de Minería*, be it to make known the names of the men related to it and who took part in its endeavors in a more or less direct fashion, be it to wipe away the dust from not a few interesting and unpublished documents that suffer within the archives from the effects of destruction, be it, specially, to present from its own perspective, an Institution that, having been the cradle of the exact, the natural, and the observational sciences in our country, has given our motherland so many eminent children, so many distinguished wise men, and has had such a favorable and immediate command over the intellectual development of our land, upon its positive progress, its glory, its well-being and its greatness.

[translated by Santiago Ramírez Castañeda.]

NOTES

[1] Ramírez, Santiago, *Datos para la historia del Colegio de Minería*, Sociedad de Ex-alumnos de la Facultad de Ingenieria, facsimile edition, Mexico, 1982.
[2] Ramírez, Santiago, *Datos para la historia del Colegio de Minería*, Ediciones de la Sociedad 'Alzate', Imprenta del Gobierno Federal en el Ex-Arzobispado, Mexico, 1890.
[3] Ramírez, Santiago, *Noticia histórica de la Riqueza Minera de México y de su actual estado de explotación*, Secretaría de Fomento, México, 1884.

ROSAURA RUIZ GUTIERREZ

GABINO BARREDA AND THE INTRODUCTION OF DARWINISM IN MEXICO: POSITIVISM AND EVOLUTION

The study of the introduction of theories of evolution in a country is interesting by itself. Further, its relation with biology in general allows the scholar to learn about the development of biology itself since evolutionism concentrates all knowledge about living beings or, at least it claims to do so.

The existence of a general life science is the essential condition for the development of the biological notion of evolution and it is also necessary for its insertion into the wider scientific culture of a country. Studies in natural history – that do not have a general conceptualization of life – are not enough to assimilate evolutionism.

To conclude whether Darwinism has been introduced at a certain historical moment it is necessary to define what is to be understood by 'Darwinism' and what is to be viewed as an 'Introduction'.

On the first issue, we must recall that Darwinism is not the only evolutionary theory of the 19th century; it must be distinguished from Lamarckism on two central questions:

First the concept of natural selection, and secondly the role of chance.

In February 1877 Pedro Noriega delivered his lecture 'Reflections on Darwin's Theory'[1] before the *Gabino Barreda Methodofile Society*. The meeting marked the beginning of the controversy about evolutionism in Mexico, but as no member of the Society was a natural historian, the debate cannot properly be considered as the introduction of Darwinism in Mexico. As Ivette Conry points out, the introduction of Darwinism into a country implies that it becomes operative within biology and that it is applied in the interpretation of biological facts.

Scholars like Moreno de los Arcos,[2] consider the discussion of Pedro Noriega's lecture at the *Society*, to be *the* introduction proper of Darwinism of Mexico. In order to reach his conclusion, Moreno considers 'Darwinists' all those who consider themselves to be such. Yet, it is necessary to search into the ideas of the participants in the discus-

S. Ramirez and R. S. Cohen (eds.), Mexican Studies in the History and Philosophy of Science, 29–48.

sion in order to know their depth of insight, their real conception of evolutionism.

The first known discussion of Darwinism in Mexico, was primarily around philosophical subjects. In 1877, Mexican positivists contemplated the need to study scientific theories, and above all, to study their methodological support. For that purpose, Gabino Barreda and his followers established the *Gabino Barreda Methodofile Society*, with the obvious aim of opposing the liberals who, by this time had succeeded in removing them from the play of political power.

In their statement of principles, positivists announced their determination to consecrate themselves to the study of social phenomena by applying scientific method; this, they believed, was the only way to solve social conflicts.

Methodologically, they proposed a profound understanding of the positive method as a criterion to justify those theories that strive to be such.

According to Barreda, Mexican history, since its independence in 1810, had been a struggle between the clergy and the army on one side, and the 'emancipated intelligences' on the other. With Juarez's reform in 1857, the church was separated from the state, favoring the "advent of a positive state"; this view was encouraged in 1867, with the definite triumph of liberalism.

Against his own Comteanism, Barreda was forced to negotiate with the liberals and to modify the classical motto from 'love, order and progress', to 'freedom, order, and progress', and to emphasize that freedom would be subordinated to order so as to promote progress.

Because of Barreda's unique set of competencies and his projects for education, he was invited by President Juarez to cooperate with the government on educational matters. Up to then, education had been in the hands of the catholic church, and at that moment, a strategy pursuing liberal goals was urgently required.

Barreda's most important project, was, unquestionably, the creation of a new educational institution: the *Escuela Nacional Preparatoria*. The fundamental aim of this project was to achieve homogeneous general knowledge among students, independently of their eventual specialization. Barreda believed that learning the positive sciences would force all educated people to think in the same way.

In this context, botany and zoology were included in the study plan of the *Escuela Nacional Preparatoria*. A general and encyclopedic culture was offered to every student. Barreda explains further that medical doctors, engineers and lawyers should be interested in discussions of theories such as evolution. For that purpose, Pedro Noriega, a student of medicine was given the task of preparing a conference on Darwin's theory.

In his paper, Noriega claimed that this theory is based upon three natural facts: inheritance, adaptation, and the struggle for survival.

From Noriega's point of view, Darwin assumes that in the beginning very simple organisms were formed; then through environmental influence, slight differences were produced. These small variations, transmitted to their progeny, would increase only if their environmental circumstances continued to function in the same way as for their ancestors. We must note that Noriega's interpretation for the need of environmental influence to last until the organisms are adapted belongs to Lamarck. Nevertheless Noriega adds to this conception the idea of natural selection. This he understands to assert that the organisms which have acquired variations that are advantageous in the struggle for survival, have greater probabilities of reproduction than those that have not.

Noriega agrees that evolution is gradual; that it is a product of the segregation of varieties and that the accumulation of small changes leads to the formation of larger categories. The way in which Darwin uses the analogy of artificial and natural selection is, for Noriega, proof enough that evolution appeals only to known natural laws that have produced analogous effects for the short period of time during which man has been capable of observing them. Noriega concludes that Darwin's theory is the only one that can explain the resemblance among species and the embryological development of each organism and the geographical distribution of plants and animals. Darwinism, moreover, agrees entirely with data obtained from extinct species. It is a *vera causa* in the sense of Herschel and Whewell.

Noriega's exposition agrees more with Haeckel's interpretation of Darwinism than with Darwin himself.

THE HAECKELIAN INTERPRETATION

Haeckel's interpretation, unfortunately is the most widespread: it simplifies and distorts Darwin's theory: Darwin admits the effects of use and disuse, but believes that the direct action of the environment is not significant; Lamarck also gives a secondary role to the production of variations by direct action of the environment since he believes it has an effect only upon simple plants and animals.[3]

It was Haeckel who emphasized the direct action of the environment in the production of variations, which is what really matters for Darwin. At the same time he refuses to accept Darwin's fundamental contribution regarding spontaneous modifications which occur with no relation whatsoever to the environment, modifications that, above all (and this is the role of chance) do not have immediate adaptive results. Their benefit depends on interaction with the environment, and their role in Darwinism is central since it is upon them that natural selection will act. Darwin explicitly points to the existence of spontaneous variations whose result could be positive, negative, or neutral: in Haeckel's version, this form of variation is overlooked. Furthermore, he claims this to be strict Darwinism; indeed, for Haeckel every variation is adaptive.

Since Noriega's lecture on Darwinism reproduces Haeckel's point of view, we can readily assume that it is was based more on the *Natural History of Creation* than on *The Origin of Species*. After Noriega's lecture, the vast majority of the association members declared themselves to be evolutionists. Luis E. Ruiz, Porfirio Parra and Manuel Flores joined in to extend their Darwinist conviction, and yet, they still fell prey to Haeckel's ideas of a continuity between Lamarckism and Darwinism, to his confusion between variation and adaptation, to a deterministic vision of evolution and to the mechanical and reductionist statement of the validity of his theory for every phenomenon in the universe.[4]

At the following meeting (February 25, 1877), Gabino Barreda disagreed with evolutionism, particularly with Darwin's theory. His most important objections were that Darwin exaggerates in his generalizations and that he has not followed the rules of scientific method in the construction of his theory. From Barreda's point of view, Darwin's evolutionism does represent progress when compared to theological cosmogonies, but this does not imply that it is correct.

Concerning the exaggerations in Darwin's generalizations, Barreda points out that, according to Darwin, only useful organs will perpetuate and that this is, apparently, a generalization derived from observations of nature, but then when useless organs are found, Darwin assumes that such organs were useful for the forefathers of the species.

Regarding the use of artificial selection as an analogy for natural selection, Barreda thought that it is a phenomenon by which certain acquired modifications can be protracted, but it cannot produce new species. Furthermore, these changes can only appear in cultured species under human care. In fact, Barreda holds a fixist position since he claimed that the true tendency of nature seems to consist of returning, through successive crossing, to an average type.

Barreda's fundamental criticism of Darwin is Darwin's lack of proofs and his effort to present a subjective conception as an objective fact. Darwin, according to Barreda, spells out the facts that he tries to explain and that nobody denies. The facts are indeed undeniable but they are still in want of an explanation.

While Barreda finds in Darwin an idea of progress, he cannot understand why Darwin does not believe in natural tendencies toward progress. Reality, to Barreda, shows that in the struggle for life, inferior organisms acquire supremacy over superior ones, since they demand more elementary living conditions and have, therefore, a greater opportunity for accommodation. Barreda sees competition as a generalized struggle while Darwin considers it as a struggle among organisms with similar requirements.

However, after Barreda's intervention, his pupils, Porfirio Parra and Luis E. Ruiz, still declared their complete agreement with evolutionism.

Porfirio Parra noted that his teacher had not yet understood the real meaning of some concepts: Barreda construed 'struggle' as real combat and not as a metaphor. He also disapproved of Barreda's claim that Darwinism states that only what is useful is inherited.

As Haeckel would, Parra gives a totally Lamarckian answer: he indicates that a descendant, while inheriting all its parent's qualities, develops those organs that are useful. In due time, these developed organs are transmitted to a new generation while those that are useless eventually disappear.

Regarding the non-existence of a continuous chain of living beings, which Barreda advanced as proof of the absence of evolution, Parra says that a linear chain cannot be expected. Rather we must anticipate something like a ramified tree; thus, we should not search for intermediate organisms in fossil remains but for organized types that concentrate those features that can only be found disseminated throughout different species.

Another member of the association, Manuel Flores, tried to prove that Darwin was in accordance with the 'positive' method. According to Flores, scientific method must state as a hypothesis, that the agents producing a cause must be, as Newton said, the *vera causa*. In other words, they must be agents really existing in nature and must be capable of producing the effects attributed to them. Following Flores, if the material element of the explanation is a *vera causa* and every law has been tested in experience, there is no need to call upon new agents or upon new laws. When this happens, the explanation no longer has a hypothetical character, but becomes an inductive truth.

The Darwinist proposals which Flores endorses are: natural selection, variety formation, inheritance and the struggle for survival.

Flores further considers natural selection to be a universal law, and suggests that it should be renamed 'law of survival or of the persistence of the fittest'. The term 'universal' is used because every phenomenon is subservient to this law: among chemical and physical phenomena, and also among those concerning the development of human works, only the fittest persist amid a progressive and changing environment. The fittest is not necessarily the most perfect, it is the most timely.

Flores, against Parra, recognizes the real existence of species and varieties that Parra rejected. Since all known species appear in a diversity of forms, Flores maintains that there must be a law ruling their conformation. This conformation is a response to specialization at the physiological level (he fails to grasp Darwin's notion of ecological specialization) by virtue of which animals are capable of living in a greater number of environments in ways "to promote the tendencies that favor them and to counteract those that are harmful".[5]

Since no one can deny that variability occurs, it can be questioned whether it leads to further perfection. Flores claims that such perfection will be achieved when complemented with the law of persistence of the fittest or of natural selection.

Based on Lamarck, Flores disputes the notion of a return to the average type as presented by Barreda. He states, rather, that the variations acquired by living beings are lost only when the bearer returns to the environment in which it was placed before the variation occurred. He also claims that the features of an animal are in large measure determined by environmental conditions and change with them; hence the possibility of talking about a 'law of inheritance'.

Flores makes explicit reference to the law of struggle for survival. In the first place, he explains that the word 'struggle', in spite of its suggestion of direct combat as Barreda noted, is a larger concept. It includes – beyond the struggle against other organisms or against environmental circumstances – certain peculiarities of organization such as color, size, or behavior by means of which confrontation is avoided in order to favor the weakest and elude danger. This cancels Barreda's doubts about the existence of organisms that, at least apparently, are not the strongest. In view of all this Flores thinks that he has proved the theoretical and methodological value of Darwinism.

After confronting his students, Barreda became aware of his total agreement with evolutionism and yet at this point he tried to convince them that Darwin had not followed the scientific method in the formulation of his theory, thus making it inadmissible. In his address, he still tried to prove that the theory of natural selection is merely a description of phenomena that cannot be explained. The description, he claims, has to satisfy no other condition than that of being a convenient representation of phenomena, not in an absolute manner but only relative to the intended objective. Moreover it must prove that the event invoked as cause exists in reality and that it can produce the attributed effect.

As long as these two conditions are not met, the hypothesis created as a basis for the theory cannot be considered as proved, regardless of its potential for presenting the facts as tied up in one sheaf, regardless of the fact that a complete description might be carried out.

The inductive or deductive proof of a causal relation between the fact that is presented as an antecedent, and the one that is assumed as its consequence is required. For a description, a simple analogy is enough, but for a theory, a constant relation of succession among the facts brought together is required. Moreover it must be such that for each case, if a fact is missing, the ensuing one would necessarily be missing. In compliance with the positive method, Barreda says that true theories should not leave out of their explanation any known fact unless the data for their application are notoriously out of reach. Even in such a case, the facts that the theory cannot embrace must not be directly opposed to the theory itself but should only be taken as unexplainable in view of the absence of required starting points. In other words, these facts cannot be retained by the theory except in a general and more or less vague manner.

Darwin, based upon the principle of *vera causa* that he learned from J.F.W. Herschel and Whewell, felt that since humanity, through artificial selection, had accomplished in a short time the production of several new varieties of cultivated species, natural selection could have transformed living beings from the condition shown by fossils to their present shape.

Herschel maintained that a complete scientific theory is one with a set of axioms from which everything can be deduced; therefore these axioms are universal natural laws. The central notion in Herschel's philosophy was the *vera causa* doctrine, which held that: "If there is a great resemblance between two phenomena and, at the same time, the cause of one of them is known then it is absurd not to recognize the action of a similar cause for the other phenomenon, even if it is not clear enough".[6]

At odds with Herschel on the *vera causa* issue, and declaring himself a rationalist (against Herschel's empiricism), Whewell admitted a Bacon–Newton type of methodology. For Whewell, the validity of a theory rests on the idea that different regions of science converge and can be derived from the same principle. He called this 'inductive confluence'. Furthermore, the causes that can be referred to by the principles are the *vera causa*, even if there is no empirical evidence for them: true causes must be adequate to explain experience instead of deriving from it.[7]

Barreda differs from the inductive position of Whewell, where every time that induction is made, that is, every time that facts are united through a general concept, we are adding a new one. Barreda argues that a general concept is far from being a new fact supplementary to the observed ones. Rather, it is obtained by the work of pure abstraction, that is, of subjective simplification of the observed facts, abolishing anything peculiar in them while leaving what they have in common.

Barreda also criticizes the use made of metaphors as a method for producing general conceptions. He says that a metaphor is frequently taken as part of the general conception. This is one of the most important criticisms of Barreda against Darwin: he points out the case of Cuvier who called the great divisions of the animal kingdom 'branches', because of their resemblance to trees; this representation is simply suggested by the attribute of subdivision. The problem arises because our minds cannot represent those facts without adding such other attributes as shape, color, size, etc. "And this, added by our mind, is precisely that which should not be part of the general conception".[8]

Barreda pushes this criticism further against concepts that function at the base of transformist ideas. For example, he states that botanists recognized certain resemblances between the covering of some fruits and the leaves of a plant and this led them to name leaves as *carpelars*. The metaphor consisted in stating that the leaves were folded from the central nerve and then joined at the edges to form the placental line. Barreda states that botanists considered this soldering a part of subjective systematization, that they have never claimed that leguminous plants had their grains exposed in a remote age and that leaves have been folding, either in the same individual or through different generations.

He also disqualifies, as a 'trick of mere description' the vision of the divided leaves as if they were originally unfolded, the bracts the sepals, and the petals and stamens as successive modifications of leaves. He states that no observation has justified that hypothesis but, on the contrary, it seems to be indicated that the real series is the inverse. Barreda accepts that there can be certain changes but adds that they are nothing but a return to normality in the cases where there has been some type of previous degradation due to exceptional influences. He leaves no doubt as to his fixist conception through which we can indeed find degeneration and then restitution, but in which "the real and objective transformations are never verified, in the sense of complication and specialization of functions or as we known it, perfection".[9]

Barreda judges the theory of the evolution of the skull from the vertebrae in the same way. In this case he more clearly opposes the concepts derived from the *Naturphilosophen*.

In fact the speculations of the *Naturphilosophen* do not require a properly evolutionist concept. Their fundamental conception is a unity of nature, according to which living beings were constructed by a supreme maker following a unique and general project; and all living forms were concrete modifications of it; the differences are ascribed to the peculiarities of their life forms. This means that all animals would be variations of an original and unique archetype. Plants would be derivations from a protoplant (*Urpflanz*), the primeval and imaginary plant of Goethe. The plant or animal archetype possesses an intrinsic equilibrium through which it only allows the development of a part of it while another wanes. The archetype modifications depend on the real circumstances in which the organism develops.

Barreda argues against the inconvenience of using metaphors that transform a simple description into a theory, since a metaphor generally is involved in the theory. He believes that there is no single description that will not forbid the inference of new facts resulting from the analogy which has been recognized, and that has given birth to the metaphor.

According to Barreda, in cases where observation does not provide enough information, the correct method is to make a hypothesis about information we do not have, or to assume that a known law can be extended to the actual facts, but it is not permissible to propose explanations that would contradict such facts. If so, our subjective creation could not become a system, and would only serve to dominate and rule the facts.

If, as Barreda holds, a hypothesis can be used in its most advantageous sense, and since it might not be a strict reflection of reality a theory that asserts objective relations between facts has to explain a majority of them and must not be contrary to any of the known empirical laws that pertain to the phenomena with which the theory deals.

After explaining his thesis of a positive methodology, Barreda tries to prove that Darwin deviates from it. To do so, he maintains that Darwin establishes the following laws:

1. The existence of a developing scale of morphological and functional complexity for living beings, from protozoa to man.
2. Progressive specialization of organs and their functions, and a conspicuous adaptation of the former to the latter that suggests the notion of purpose.
3. Division into two very similar kingdoms, maybe even identical: but they achieve, as they perfect themselves, a radical deviation that cancels any possibility of confusion.
4. Strength of vital resistance, active as well as passive, but in an inverted sense of that expressed above, that allows organized beings to endure the unfavorable conditions of their habitat, to reproduce abundantly, and to mend traumatic injuries to a point of almost total restoration, faster when the organization and functions of the living being are simpler.

Barreda elaborates on his critique of the theological and pre-Darwinian explanations that pretend to explain the above by asserting they imply a superior intelligence that has preestablished a motive for each function, and a purposeful adaptation of each living being to the environment where it dwells. This kind of theory, Barreda points out: "cannot be used but to sanction what we already know, and are incapable of inferring or foreseeing what is unknown".[10] Moreover, he considers such theories adverse because they lead to positions where man has nothing to do because the world is the best possible.

In spite of all his criticism of Darwin, Barreda recognizes that: "... he has tried to put the issue upon the real ground of science"[11] and recognizes (yielding to the overwhelming defense of his students) that Darwinism is at no point contrary to his idea of science. Hence the problem is to know whether the laws and facts that Darwin holds are capable of producing the effects that are attributed to them. These facts are:

1. The evident, even if sometimes unexplainable, action of the environment in which a being lives to produce in that being, patient but unceasing and progressive adjustments.
2. The possibility of attesting to sometimes spontaneous, accidental, morphological changes in some individuals that differentiate them from others of the same species, such as the absence of an organ, the presence of another not normally present, the fusion of two or more, etc.
3. The ability to transmit to offspring one or more alterations, as well as the ability to establish them in certain circumstances, and even to exaggerate the resulting differences to a highly noticeable degree.

From these facts evolutionism tries to explain the main biological laws that Barreda has asserted. At this point, Barreda reduced the facts that Darwin's theory tries to explain to inheritance and variation. Even if it is true that throughout his life Darwin was concerned to explain the way variations are produced and inherited, he never claimed to have a decisive explanation, but merely a 'provisional hypothesis', like pangenesis. The facts that Darwin explains are those that follow variation. He takes it to be a fact that all organisms are different and that such differences define the organisms as more or less fit. This leads Darwin to the struggle for survival that compels natural selection to eliminate some individuals and to preserve others, in general, different from their forerunners.

Barreda holds that of the four laws the first and second (increasing complexity and adaptation), are the only ones that Darwin tries to show deductively; of the two others, he claims that they contradict the theory because they cannot be presented as consequences of the struggle for life, of natural selection, of the division into two kingdoms, or of the major vital ability of inferior beings. It is important to reiterate that these laws are stated by Barreda, not by Darwin, and if indeed the third and fourth are conflicting with the theory, the theory of evolution by natural selection remains untouched.

As incontrovertible proof of the mendacity of the claim that the struggle for existence is a motive for an improvement in life conditions, and thereby of a better adaptation, Barreda offers the fact that inferior organisms are the survivors and superior beings offer less resistance to the causes of death. Moreover, he states that many of the simpler organisms, like small lizards, crustaceans, and all plants, can restore one or more members of their body without missing a single detail of their complex organization.

If the improvement of life conditions were a result of the struggle for life, Barreda writes, the organic kingdom would have never pulled away from the inferior levels, that, being capable of survival under conditions that permit the life of higher beings, could also adapt to circumstances under which these cannot survive, much less prosper.

To paraphrase Barreda, if struggle for survival were real, the winners, no doubt, would have been the inferior organisms; in that case the superior animals including man would never have appeared.

> The law of tenacity of life, thus, cannot be explained by Darwin's theory. It is contrary to it because here [life] would tend to the disappearance of the higher beings, as being less fit for struggle (even if their existence had already started) rather than to their perfection.[12]

Regarding the third law (division of organisms into two kingdoms), Barreda claims that it cannot be explained by Darwinism and that in reality it was no concern to Darwin. He bypasses the organic scale and the increasing complication, when he asserts: "... these are the facts that the transformist theory believes it explains appropriately".[13]

According to Barreda, these phenomena would have to be understood by Darwin by calling upon the laws of natural selection, variety formation, inheritance and the struggle for survival that have been set forth by Manuel Flores. Yet Barreda does not agree with Flores in equating natural selection with survival of the fittest. He argues that no transformist could do so, because natural selection gradually prepares and determines the advent of the fittest, while survival applies when the fittest already exists.

Regarding inheritance and the possibilities of modification, Barreda captures the essential part of the Darwinian thesis (about inheritance and the possibilities of modification) and points out that the differences that might accidentally appear either alter the individual, allowing it to satisfy needs more efficiently, or, on the contrary, are detrimental. In the latter case, the carriers of the change tend to disappear, while in the former the result is the stability of the species, but modified.

Undoubtedly, he continues, adaptation, variation, and the struggle for survival are real facts. With Linnaeus, he accepts the creation of new varieties as a phenomenon restricted to the species level, which does not assume evolution. It is evident, he says, that nothing can exist if it does not satisfy the conditions for its existence. He further admits that individuals may present variations that can be transmitted and perpetuated by inheritance until a new variety is created, and that there is some sort of struggle or competition between the environment and the beings that inhabit it as well as among these beings. The result is that individuals who do not possess the means to struggle advantageously tend to vanish. Barreda contests that such facts can produce the evolution of life; what is impossible to admit is that: "... these (facts) are enough to give an account of the phenomena and are thought of as their effect".[14]

Variations have limits, so that the profound differences in organization cannot be explained by them, not even with the aid of time or inherited transmission. He makes clear that he does not want an experimental proof of the facts that must have taken place; what he asks for is proof based on observation and experience.

What must be proved through observation – and if observation is not enough, through experimentation – is the fundamental law upon which the explanation rests. However, he insists that transformists have never been able to prove that a species changes into another under the influence of natural or artificial conditions, least of all in the case of a genus, and "... consequently, the similarity is not permissible or rather it is self-defeating, because it misses the essential condition".[15]

It is true, Barreda recognizes, that individuals show morphological or functional differences, but they are nothing more than simple accidents whereas the fundamental characteristics of the species are never lost.

It is clear that if Barreda does not admit the creation of new species, it is more difficult for him to accept changes at the genus or class level; he holds that a bird will always be a bird and that a mammal will always be a mammal.

Congenital monstrosities have been used as proof that inherited modifications lead to the formation of varieties, but as Barreda explains, even if the existence of such changes is undeniable, he considers them deficiencies that in no way could engender new species. "What scientific support", he asks, "can be claimed by the assertion that an inferiôr species can, because of an accidental development of one or more organs, form part of another in a higher step of the scale?"[16]

In addition, Barreda indicates that hypertrophy and atrophy are the only natural or artificial means of diversification, but in spite of their great power, they are not enough to create something new:

> The environment by itself, as Lamarck and Darwin recognized, is not capable of producing those radical changes, and thus the former called upon a metaphysical causality, and the latter thought it necessary to appeal to accidental variations.[17]

Isolation for reproduction is a key and indispensable element for Darwinism. If varieties are not separated to reproduce, new species cannot be formed, since interbreeding would abolish the existing differences.

This important point is also disputed by Barreda, who claims that it is of no consequence how much time passes since the inception of a variety, because no matter how big the morphological differences that characterize and distinguish it from the original are, the ability for fecundation remains.

Darwin does not claim that among varieties there is a decline in fertility since he considers them to be emerging species, not yet separated species; instead, he asserts, varieties may remain geographically or ecologically separated, stimulating reproductive isolation. Whence he considers geographic isolation as more efficient, but accepts that in an unbounded area, reproductive segregation may occur.

Barreda's last criticism of Darwin's transformism claims something that others had already stated: the personification of natural selection. He disapproves of Darwin's representation of nature:

> Diligent and solicitous mother who unceasingly and with perfect and inexhaustible dexterity endeavors to provide for the welfare of the beings it nourishes and give them everything they need to fulfill the goals for which they were called upon.[18]

Barreda's refutation is not based, as he says, upon methodology alone. That, in Barreda's opinion, is not scientific. He also opposes transformism as such, both Lamarckian and Darwinian. He accepts, with

Lamarck, that organs can strengthen or can weaken through function or misfunction. He accepts, with Darwin, competition among living beings, but he totally rejects the possibility that changes are continuous and could lead, together with competition, to an increasing transformation of life.

We can conclude that Barreda aligns himself with fixism:

> ... The real tendency of nature seems to be the turning back, through successive breeding, to which the mixture of individuals in their natural state, naturally predisposes, towards an average type.[19]

All of Barreda's disciples defend evolutionism, not only Darwin's, but also Lamarck's. They are evolutionists. However, their knowledge about the two theories is somewhat crude, possibly because not one of them is a naturalist and possibly because they all know Darwin through Haeckel. This is quite evident from the way in which all of them (Noriega, Parra and Flores) adjust the two evolutionary theories.

Pedro Noriega, the first to expound Darwinism to the *Methodological Society*, started his lecture by describing the way in which Darwin said that variations are produced. Noriega only states the environmental influence, one of the Lamarckian ways of evolution, and does not mention variations that appear spontaneously, and that do not have a direct relation to the environment, a fundamental proposition of Darwinism.

In a similar way, Porfirio Parra and Manuel Flores focus their defense of Darwinism on explanation of the variations due to direct action of the environment, by use and misuse, and on its incorporation through inheritance into the offspring. This implies ignorance of Darwin's contribution to evolutionism because even if he admits the existence of such variations, he believes that they can produce evolution since they are adaptive changes. Nevertheless, the novelty of Darwin's theory is that random variations may lead to evolution only when favored by natural selection, and Darwin never worries, as Haeckel would, about searching for confirmation of the inheritance of acquired characters. Some of Barreda's criticism is based on his fundamentally anti-evolutionist conception, some on his positivist philosophy, and some on his incomplete understanding of Darwin's theory.

Barreda pretends to analyze Darwinism as a philosopher. He tries to judge its logic and does not confront it with reality, except in particular cases. Nevertheless, we can recognize in his readings of Darwin a deeper knowledge than that of other positivists in the *Methodological Society.*

It is by virtue of this philosophical analysis that Barreda considers Lamarck's theory much more acceptable, since it generalizes observation of embryological development to species development, that is, it proposes evolution by induction. Barreda, because of his idea of adaptation, admits that organs can develop or degenerate according to their use or misuse, but while accepting the possibility of inheritance of modifications, he cannot accept that they can lead to evolution of the species. Rather he claims, that there is a tendency to return to the average type when the modifications in subsequent breedings are lost. This, in addition, requires a notion of mixed inheritance.

Some of Barreda's criticism arises from his lack of knowledge or of understanding. For example, he points out that according to Darwin only useful organs will propagate indefinitely, while Darwin considers gradual changes as the basis for his theory, and the same thing does not apply to his note on incomplete organs.

I have pointed out that Barreda does not know Darwin's solution to the problem of avoiding the crossing of species in the process of species formation, that is, of reproductive isolation accomplished by geographic or ecologic separation.

His critique about monstrosities not being capable of producing new species is something that Darwin would have agreed to, because he claims that monsters, organisms with rough changes, are not feasible in general or never reach a reproductive age, and thus, their characteristics are not perpetuated.

Another mistake in Barreda's interpretation is due to the fact that he understands 'struggle for survival' as a combat in which one of the contestants has to terminate. For Darwin the struggle for survival is much more than that. It includes the struggle against physical elements, extremely subtle competitive relations, that are stronger among organisms that are taxonomically alike than between distant species like the ones Barreda notes when he talks of the struggle between superior and inferior organisms.

This kind of struggle, of utmost importance in Darwinism, does not yield the necessary result of extinction of the competitors. Rather, Darwin proposes that on occasion, shunning competition can be a mechanism that favors species or populations in their struggle for life. In *The Origin of Species* it is stated that varieties of a species can occupy different places in nature's economy because of the possibility of morphological or behavioral divergence, resulting in the diversity we now know.

Depredation and parasitism are also forms of struggle, yet Darwin observes that predator populations do not annihilate the prey populations. Instead, he sees them as a control on the size of populations.

Of the four laws that Barreda attributes to Darwin, it must be pointed out, first, that Darwin does not state such laws. Moreover, he never proposed proving the existence of a scale (first law) because it is one of the concepts which he quickly rejects. He proposes something more like a tree in which many groups are at the same level, as against a scale in which species follow one another in a continuous line with no two groups, species or individuals occupying the same level.

Regarding a finalist tendency to specialization Darwin indeed tried to explain the causes of an increase in specialization of organs and its functions, as well as of organisms and species. He found such explanation in the divergence of characters favored by natural selection, generating a growing specialization.

However, we do not find a finality, since living beings will adapt to real circumstances that change at each moment so that a plan that contemplates such environmental changes cannot exist. Darwin, in this sense, follows Charles Lyell and agrees, therefore, with the unpredictable and constant character of such changes. Furthermore, there is an explicit rejection of theories, such as Lamarck's, that proclaim the existence of a tendency towards the increase of complexity.

Barreda's law of the division into two analogue kingdoms is not Darwin's concern. Even if it can be considered the result of gradual evolution, Darwin is only trying to explain the formation of species and does not think in larger categories. As to the fourth law in which Barreda refers to a biological superiority of inferior organisms, Darwin never states such an idea.

Some authors who approved of transformism, pointed to Darwin's somewhat finalist character of natural selection. This accusation of finalism is related to the personification of nature about which Darwin argued:

> It is hard to avoid the personification of the term nature, but I understand by nature only the global action and product of the

> multiple natural laws, and by laws I understand a sequence of events, as we actually verify them.[20]

Wallace attributes this personification to the comparison between natural and artificial selection. This ushers in a misunderstanding, since in artificial selection, men seek advantageous modifications; it is mankind who isolates varieties for reproduction, and favors those which interest him most. Natural selection itself causes such effects. Darwin contemplates natural selection in many ways: as an agent, as a process, as a result, and sometimes he refers to it as to an entity which directs evolution:

> It may be said that natural selection is daily and hourly scrutinizing, throughout the world, every variation, even the slightest; rejecting that which is bad, preserving and adding up all that is good; silently and insensibly working, whenever and wherever opportunity offers, at the improvement of each organic being in relation to its organic and inorganic conditions of life.[21]

This is not an attempt to prove that Darwin's transformist theory is finalist; it is definitely not, but we agree with Limoges[22] when he points out that Darwin, while trying to explain his thesis in a simple and pedagogical manner, lapses into confusion on artificial selection.

However, the theory results in anti-theology, because it does not accept creation and proposes an unforeseeable development that depends on a variety of factors, each with its own movement and with no direct relation to the others. It is clear that for Darwin, evolution is contingent, for its result could have been different.

Barreda's analysis, finally, follows from his anti-evolutionist position; he believes in the stability of species, with his claim of a tendency to revert to the original type, and while he accepts that variations are real, he rejects the notion that they lead to the formation of new species, no matter how drastic or negligible they might be. This also explains his partial agreement with Lamarckism, which he considers to explain modifications as a result of relations with the environment, and yet he will not accept that these changes lead to evolution.

Barreda holds that according to the positive method a theory must be drawn by induction from direct observations and by experiments which will allow generalization of what has been observed and that a theory cannot be construed (as in Darwin's case) in such a way that it is confirmed afterwards with the use of *a posteriori* metaphors and experiments.

From this point of view, Barreda is right: Darwinism is not a theory that relies on observations and experiments alone, with no preconceived ideas. On the contrary, it originates in a conceptual interpretation of what has been observed. To achieve this aim, it appeals to other theories (not all of them biological), and later seeks, in experiments, confrontation with reality.

In summary, the controversy that took place at the *Methodological Society*, with all its limitations, undoubtedly played an important role in the diffusion of Darwinism, but as was pointed out at the beginning, no member of the group was a naturalist, and so the introduction of evolutionism to biology did not take place in Mexico at the end of the 19th century when biology was still in its inception.

NOTES

[1] Noriega, P., 'Consideraciones sobre la teoria de Darwin', *Anales de la Asociacion Metodofila Gabino Barreda*, Imprenta del Comercio de Dublan y Chavez, Mexico, 1877, p. 98.

[2] Moreno de los Arcos, Roberto, *La Polemica del Darwinismo en Mexico*, UNAM, in press.

[3] "Ici, il devient nécessaire de m'expliquer sur le sens que j'attache à ces expressions: *Les circonstances influent sur la forme et l'organisation des animaux*, . . . assurement, si l'on prenait ces expressions à la lettre, on n'attribuérait une erreur; car quelques que puissent être les circonstances, elles n'operent directement sur la forme et sur l'organisation des animaux aucune modification quelconque". Lamarck, *Philosophie zoologique*, F. Savy, Paris, 1873.

[4] Cf. *Annales de la asociacion metodofila Gabino Barreda*, pp. 111–112 and 125–136.

[5] Barreda, G., in *Anales de la asociacion metodofila Gabino Barreda*, p. 136.

[6] Herschel, J.F.W., 'Preliminary Discourse on the Study of Natural Philosophy', in Ruse, M., *The Darwinian Revolution*, Chicago University Press, Chicago, 1949.

[7] Whewell, W., *Philosophy of the Inductive Sciences*, London, 1840, 2: 442.

[8] Barreda, G., *Anales de la asociacion metodofila Gabino Barreda*, p. 145.

[9] *Ibid.*, p. 149.

[10] *Ibid.*, p. 159.

[11] Barreda, *loc. cit.*

[12] *Ibid.*, *loc. cit.*

[13] *Ibid.*, p. 163.

[14] *Ibid.*, p. 166.

[15] *Ibid.*, *loc. cit.*

[16] *Ibid.*, p. 169.

[17] *Ibid.*, p. 170.

[18] *Ibid.*, p. 152.
[19] *Anales de la asociacion metodofila Gabino Barreda*, p. 104.
[20] Quoted in Limoges, G., *La sélection naturelle – Etude sur la première constitution d'un concept (1837–1859)*, Puf, Paris, 1970.
[21] Darwin, *The Origin of Species*, p. 140.
[22] Limoges, *op. cit.*

SANTIAGO RAMIREZ RUIZ

PSYCHOANALYSIS AND MARXISM

Nerval recalls poetically, in *Aurelia*, the challenge facing humankind ever since the beginning of times: "*Le rêve est une seconde vie*".

Ever since Joseph, dreams were that parcel of life wherein men met the gods or the cruel and unmerciful world where Oedipus was condemned. In both cases, a dream serves as the key to behold other times: in the former, it is the announcement of the future; for the latter, the clue where the sense of the past is to be found. However, in either case we find a breach, in either case dreams escape every human control and as for Nerval, a dream is alien and strange, it is another life.

In more general terms, the problem posed by Nerval is but another sign of an age-old problem that has been – more or less fortunately – solved, elsewhere, by Freud when he uncovered a unity for the manifold: when he thought of man as a unitary being yet he preserved the multiplicity of human presence.

Nerval wrote: "*Le rêve est une seconde vie*" and such seems to be the human drama: a two-fold drama in which, on the one hand, man is torn between two modes of existence: awareness and fantasy: a drama; on the other, that places man in a cleavage where the world, in which the major part of life takes place, cannot be understood: the world of dreams, the hallowed world of Joseph or the merciless and unrelenting world that doomed Oedipus. It is, indeed, a second life that is not only at odds with the first but that transcends man itself.

This replication had been solved, before Freud, on other levels. Thus, in the *Phenomenology of Mind*, Hegel reaches the death of God. A death that reconciles the world of men with the world of gods; the ruin of God in two senses: as the humanization of divinity – Christ – and as the consecration of the human. Marx is capable of overcoming the apparent duplication of life and consciousness by showing the latter does not determine life but life determines consciousness.

Freud's first task is to show that the dissension between the dreamed and the 'real' world, the divorce between perception and imagination, the difference between consciousness and unconsciousness, are but de-

S. Ramirez and R. S. Cohen (eds.), Mexican Studies in the History and Philosophy of Science, 49–57.

ceptive and illusive. Freud strives towards recovery of the primary unity that is manifested through this duplication allowing for the appropriation of that which originally showed itself as the Other.

If Hegel has spiritualized man to undergo his divinity and Marx has shown the way through which man must trek to recuperate life, Freud has revealed the secret for the repossession of the 'second life'.

We encounter, hence, in Freud, a discovery and consequently new possibilities for recovering the world in general and the human 'essence' in particular.

Furthermore, this discovery is a scientific discovery and is not, as has often been held, a dogmatic or ideological cover-up. If psychoanalysis is frequently depicted as a dogma, as a closed and rigid entity, as an 'orthodoxy', we must make clear that, even if it is a *corpus*, it is not a "proposition that asserts itself as undeniable", and we must clarify that orthodoxy does not mean, or must not mean, "dogmatic uprightness or compliance with dogma". Rather, psychoanalysis is a flexible theory, resilient and exuberant. To claim the contrary would be, for example, to deny that medicine lacks an evolution: it is obvious that the medicine of Hippocrates, under the omens of the oath and of the Delphic Oracle, is not the same as medieval alchemy; that one medicine is Harvey's, pregnant with anatomy, discovering the circulation of blood while another is Pasteur's bacterial medicine or Fleming's antibiotics, not to mention present radioactive medicine.

Orthodoxy – and we shall return to the point later – is, as Lukacs puts it, a question to be referred exclusively to method.

And Freud's method – in spite of its apparent inconsistency – is definitely a scientific method.

This calls the question, What is science?

Above all, science is history.

Of course the claim does not belong to Freud, or even to Marx. Its originality is due to Hegel.

Thus, in Freud as in Marx, in Freud as in Darwin, history is the instrument authorizing us to discover and is, in itself a discovery: Darwin's evolution, Marx's class struggle, Freud's childhood: Infancy is Fate.

Moreover, history works as the element of explanation in Freud, Marx or Darwin, not as a linear or continuous history, not the chronicle of an uninterrupted accumulation. On the contrary, history is made by bounds and rifts. It is movement, the development of the contradiction between the manifold and the unique: it is the unfolding of the unity in its different forms and the recovery of the unity of the multiplicity of forms. Darwin's evolution is a prancing process generated by the differences between the organs and their functions, Marx's is the history of the conflagrations between the various classes and the boundaries are revolutions. In Freud, the axis is the confrontation between the principle of reality and the pleasure principle, between the primary process and the secondary process, between awareness and dream.

Conduct is diverse and its significance is ambiguous. Freud's chore is to uncover the essential unity lying in conduct.

Secondly, science is a transformation and such is the content of Marx's eleventh thesis on Feuerbach when he states that the purpose is now to transform the world.

Negatively put, science is not an interpretation of the 'real', a recognition of existence or an explanation of what is given. Freudian theory, despite its title, *The Interpretation of Dreams*, is not, in the sense noted, properly, an 'interpretation'.

It is rather a transformation, a reproduction, a becoming another (*von anderes*). It is a reproduction to produce again in the following sense:

For Marx – as for Kant and Hegel before him – the object of contemplation, the object in front of which the dazzled subject is marveled (*Objekt*), reality as a chaotic whole (*Realität*) must be distinguished from the object 'in objectivity', from the result of human activity (*Gegenstand*): from effective reality as the 'synthesis of multiple determinations' (*Wirklichkeit*).

The above derives from the etymology of the word *Wirklichkeit* itself: its origin is *wirk*: to do, so that effective reality is that which is done. It is what has been produced from a prime-matter (*Realität*), from present reality, through the action of a subject: an action that, by the way, simultaneously produces its subject that is not now an assumption but a product.

Science is the theoretical moment, the thought moment of this doing, of this reproduction of the real: it is the reproduction of the real through thought. Let us stop, with Heidegger, with the notion of theory.

In its fundamental connotation, theory means 'to rise to contemplation', 'to remain close (*bei*) to the object', to put it at the disposal of the subject. Thus there is no appropriation but of what has been laid at the subject's disposal.

Science is also an opening. It is the opening of a dominion of being, an opening that must mean 'to put in a position to be appropriated'.

Such is Freudian science.

Far from being the opening of a continent containing a mystery, far from proposing a new 'irrationality' for men, it slowly outlines something that cannot be outlined (*unumgänglich*) and places it at men's ordering. This *unumgängliche*, whose naming solves nothing, this 'soul', appears in Freudian theory with a content: history, past, infancy, binary or triangular situation, a knot of conflicts. Thus, Freudian science is the opening to a dominion of penetrating, outlining, the so-called 'unconsciousness'.

Freud not only lays out the 'soul' in its objectivity, transforming it into the unconscious, but places the subject, man, in a position to grasp it and to grasp his own history, no more as a sequence of chaotic episodes but as the synthesis of the various determinations of the subject.

Freud not only opens a continent but gives man the weapons to conquer it .

As for Marx every historical moment is – in its universal determination – class struggle, and the particular moment is a knot of contradictions that can be reduced, in the last instance, to class struggle and to its placement in the productive process, so for Freud a particular subject is nothing more than a knot of conflicts that can be grouped – *grosso modo* – in three nuclei that are nothing but moments, split instances, of an original unit: *id*, *ego* and *super-ego*.

These determined elements are not the result of Freudian theory but abstract points of departure. From these three unilateral and abstract elements, Freud can carry but the synthesis that allows him to reproduce that originally chaotic whole that is the subject.

Hence Freud has opened man towards his unconscious and his personal history, to be able to contemplate that history, and he has shown unconsciousness by (*bei*) man.

But if Freudian theory stopped here, it would be a mere therapeutic procedure, a method for healing. Freud's object, the 'content' of his continent cannot be confined to that of something else. Freudian theory, Freudian science, is not an appendix to medicine in the same measure as his interpretations of Leonardo or of Michelangelo's *Moses* are not a sequel of aesthetics: Freudian science is neither anthropology or sociology as we are led to believe by those who have not yet understood Freud, as Wilhelm Reich taught Erich Fromm. It cannot be anything of the sort because its object is absolutely unprecedented. Freud is not a witch doctor nor a witch in medicine. Such misconceptions are the burden that psychoanalysis has shouldered when it has been transformed into an element of the dominant ideology, a fact that cannot be blamed on Freud but upon those of his followers who have transformed psychoanalysis into a *modus vivendi.*

Freud's object cannot be referred to anything else, and Freudian science, psychoanalysis, hence, cannot be considered as subservient to any other. Freud tries, from the knowledge of individual histories, to reproduce culture. Thus, individuals are the elements of a concrete totality even if their sub-total is not the concrete totality of which they are but a part. Society is not the union or joining of its elements but the synthesis of them. In Freudian terms, individuals will not explain culture but culture shall give an account of individuals that cannot live as psychic Robinson Crusoe's. Man is – as Hegel claimed – relations. Later, Fairbairn will explain "internal psychic structures in terms of object relations". Freud tried to explain – his success being irrelevant – cultural formations in terms of the history of individuals. Yet, his explanation is not a metaphysical explanation searching for something common to everyone, nor an induction that, taking hold of the unconscious of a particular individual, elicits the traits of culture. Freud tries, on the contrary, as we have just claimed, to explain the individual from culture, to retrace his itinerary in the opposite sense, to reproduce culture itself. That is, starting from culture-*realität*, he traverses abstractions – in his case individuals – to encounter, again, his point of departure, but this time, with knowledge about it: he arrives, at last, to culture as a *Gegenstand*, as *wirklich.* In this sense, Freudian theory is not an interpretation of culture but a transformation of it to put it at the service of man.

In this manner, Freud overcomes unconsciousness: with Freud, unconsciousness disappears as such. With Freud, the unconscious is not interpreted in a new or in a different way but is transformed into its opposite: man acquires a consciousness about unconsciousness, he recovers something that had hidden for a long time and that had dominated him.

But history does not appear in science as an explanatory element alone, as the axis of a scientific theory but it itself is the explanation of science. Otherwise we could not distinguish between science and poetry, between Freud and Rimbaud. We would have no way to explain the difference between Freud's scientific abstractions and ideological abstractions, pre-scientific notions spread out through those handbooks of dream interpretation.

The substantial difference between poetic thought and scientific thought – whose similarities are much more profound than those that can be given in a subjective fashion – is the history peculiar to each. Art and poetry, from a certain point of view, are immune to time's flow. Science, on the contrary, is deeply linked to the history of human society. Thus, to understand a scientific theory we cannot overlook the particular history of that science and of the context in which it develops. However, we cannot write, here, the history of psychoanalysis. We only will point out some elements that would be useful for such a task.

We conceive history of science from two basic facts:

First, science is not formed through a slow hoarding process but rather through large revolutions from whose ruins a new explanation flowers, denying everything past. We think it is clear that with Freud we face the systematic destruction of every former theory of the 'soul'.

Second, we believe that scientific theories do not appear, in their initial moments, as such but as concepts around which and from which scientific theories are structured. These concepts are not, in their turn, arbitrary products but must be explained under the light of the problems that the culture of the times allows to be posed and solved. The history of science, hence, must be related to the cultural or ideological junctures that make the appearance of a concept possible as the answer to a problem.

We shall try, in what follows, to trace some of the concepts proposed by Freud and to account for the manner in which, from them, psychoanalytic science can be construed. We do not claim, at this point, to establish those junctures that have made these concepts possible or the meaning that they have for the surmounting of earlier explanations.

Therefore, we must bear in mind that Freud's thought suffers severe mutations from the eighth decade of the 19th century to the third of the 20th: between *Hysteria* (1888), the *Project for a Scientific Psychology* (1895), his correspondence with Fliess and his last epistolary work, there is a ravine apparently burdened with contradictions. Few people know in analytic and synthetic form the work of Freud. In our (Mexican) milieu, the number is still narrower, including the technicians of the *modus vivendi*: our psychoanalysts. Freud, fortunately, had no time to revise his own work, in order to smooth his own conflicts. Fortunately, since that allowed that we were bequeathed from his exile in London some of his most brilliant works: *Analysis Terminable and Interminable* (1937), *Constructions in Analysis*, and *Moses and Monotheism*. In a very brief sketch, we shall try to outline the fundamental moments of Freud's work.

Struck by the case of Anna O., he publishes his first clinical cases: Emmy, Elizabeth, Lucy and Kathy. At that time, he deals with an iconic language. He assumed that a unique trauma was sufficient to produce a symptom. A toad frightened Emmy when it leaped while she picked up a stone, she started babbling. Until then and for a long time afterwards, it was ignored that the trauma was not such but rather a simple episode which could be taken as an analogy to a sign loaded with a complex and enduring significance. In a systematic way, for many years, the signs were confused with the significances and it was thought that with the revelation of the sign, without touching the significance, the illness disappeared. To a vast measure, this discovery, exact, but particularly partial, gave birth to the phrase that defined cure as 'making unconsciousness conscious'. Something had not been taken into account, something that was given within the patient–analyst relation: the relation itself. On the other hand, this shortcoming is not only present in the psychoanalytic relation but also in the medical, the pedagogical and the political ones. In 1905, stemming from the outcome of Dora's treatment and the analysis of 'Mr. K's cigar', it becomes clear to Freud that he himself can be represented because of the coincidence of his own smoking habit with that of a real character of Dora's life: Mr. K. Here was a substantial discovery, axis of the therapeutic method and of subsequent theoretical considerations: transfer. Around this subject, we find *The Dynamics of Transference* (1912), *Observations on Transference-Love* (1914), etc.

A new substantial moment: there are individuals who remember so as to avoid repetition (obsession), others who repeat so as not to remember (mania and hysteria). *Remembering, Repeating and Working-Through* (1914) is a practical and theoretical pillar of psychoanalysis. The last repetition in the interpersonal relation is that appearing in transference: the manner of greeting, of saying good-bye, of seduction, of joy, are nothing but reiterations of remote pasts. Conduct is a piece of history. Strange phrase that has cost so much to accept into psychological matters and so little, say in linguistics. We have no idea of the power it still has, in our conceptual distortions about 'free will'.

The last and most polemic moment in the inception of Freudian theory is that of the 'compulsion to repetition'. While being unable to explain the inertia against future change, he is forced to introduce precisely that, a principle of regressive inertia, 'death instinct'. Doubts about it derive from the following: Does it really exist? Or could we claim that a child found frustrating and evil objects, but nevertheless objects. He prefers these because the opposite, unlearned, would be an annihilating solitude?

Briefly, maybe partially, from our point of view, these are the determining moments. Not only are episodes and significances repeated but also a rhythm and a timing, a melody and the dissonances with which and in which they happened.

We can now enumerate the fundamental postulates of Freud's work:

1. Conduct is motivated. We shall not lapse into the examples derived from experimental psychology, anthropology and sociology. Hunt, Margaret Mead and Kardiner are but three notorious examples of the above.
2. Most of the motivations of conduct are unconscious.
3. Most of the unconscious motivations that define conduct are begot in childhood. Oral, genital, procreative, nourishing conducts.
4. The motivations of conduct bred in childhood tend to be repeated. 'Compulsion towards repetition', 'death instinct', in colloquial language, "everyone gets what he looks for", or further, everyone puts the milk he can find into his own cup of coffee. We can add: "When life does not offer to everyone his corresponding other, everyone searches, exactingly, meticulously, to modify the conditions surrounding him so as to make them fit his internal needs".

From these postulates, basic in our opinion and with unlimited implications, stem the obstacles to change. It is necessary to go through a protracted and permanent transformation of the external world to achieve transformations in the internal one, individually as well as socially.

We conclude our remarks, apologizing to our readers for having to bear with us the difficulties of a complicated dialogue, not only because of the enigmas derived from it but because of our intention to establish links between psychoanalysis and Marxism and because of the difficult historical conditions facing the author. We have tried to give a sketch, not only of the conciliation between philosophy and psychoanalysis, but also between generations. About this concoction we plead absolutely guilty.

But we still believe, and firmly believe that only through a comprehension of psychoanalysis that takes into account the multiplicity of values converging in it, only by overcoming the natural curiosity that will merely look, only then are we going to contemplate psychoanalysis as a response to the problem of the essential unity of man, of human conduct and of human life.

[translated by Santiago Ramírez Castañeda]

REFERENCES

Freud, S., *The Dynamics of Transference*, Standard Edition, 12, Hogarth Press, 1974.
Freud, S., *Observations on Transference-Love*, Standard Edition, 12, Hogarth Press, 1974.
Freud, S., *Analysis Terminable and Interminable*, Standard Edition, 23, Hogarth Press, 1974.
Hegel, G. W. F., *Phänomenologie des Geistes*, Werke, 3, Suhrkamp Verlag, 1970.
Heidegger, M., 'Wissenschaft und Besinnung', in *Vorträge und Aufsätze*, 1, Neske, 1967.
Marx, K., *Thesen über Feuerbach*, Werke, 3, Dietz Verlag, Berlin, 1973.

CARLOS PEREYRA

FUNCTIONAL EXPLANATIONS IN HISTORY

During the development of the theory known as *historical materialism*, concern about the precise nature of the explanations that such a theory makes possible has been extremely meager. There has been some discussion of the idea that we are dealing with *dialectical* explanations about the workings of society as well as about the observable transformations in such a way of operation. Emphasis is on the dialectical quality of the explanations furnished by historical materialism, and it is common to make reference to the notion of *totality* and to the role that is played in the latter by *contradictions* as the ultimate foundation for change. Whatever might be the possibility of conferring a larger content to the claim that explanations derived from historical materialism have a distinctive peculiarity ('dialectical'), the truth is that with it the need to relate them to other explanatory procedures used in the social sciences has not been eliminated. "... in particular, it can be asked whether historical materialism is a causal or functional theory, or whether it can be both" (McLennan, 1981, p. 45).

In recent years, certain researchers in the tradition of historical materialism have worked on the assumption that "materialist explanations are typically not causal explanations ... they are teleological explanations" (Wood, 1981, p. 104). That assumption has also been formulated in a way that would avoid the undesirable echoes of the word *teleological* and establish more clearly the claimed feature of explanations: "the basic explanations of historical materialism are what we have called functional explanations" (G.A. Cohen, 1983, p. 111). In these proposals we could see a slow effect of equivalent quests in other theoretical traditions, since more than two decades ago it was said "for some years now, the researchers in social sciences argue about the convenience of the so-called 'functional analysis' in their disciplines" (E. Nagel, 1961). Of course, we are trying to invoke these antecedents in order to disqualify those proposals, and the fact that anthropologists and sociologists debated the pertinence of functional explanation years before such

S. Ramirez and R. S. Cohen (eds.), Mexican Studies in the History and Philosophy of Science, 59–70.

a discussion was deployed within the theoretical horizon of historical materialism has no importance.

What, then, is a functional explanation? Maybe the first step toward an answer is to distinguish between the description of a thing's function and the explanation of the presence of something because of its function. Hence the validity of Cohen's criticism of Merton since the latter "never satisfactorily distinguishes between explaining something by reference to its function(s) (functional explanation proper) and explaining the function(s) of something" (Cohen, 1980, p. 283). Even if the thesis of Cohen is correct, its wording is ambiguous for he uses the term *explanation* in both cases thus obscuring the distinction whose absence is reproached in Merton. The word *explain* appears in both cases because Cohen refuses to admit that in scientific discourse the term must be reserved for those propositions that answer the question, why? Such a refusal is supported by the multi-meanings that the word has in ordinary language where it is possible to demand, "explain the rules of chess' as stated by someone who wants to know what these rules are and *not* why they are thus. It is equally feasible to ask for an 'explanation' of what is going on in Central America when one wishes to have information and not when searching for the causes of what is going on there. The ambiguity of the word *explain* in everyday language, nevertheless, should not serve to muster *equivocality* into scientific discourse where the term is always referred to the question, why? Thus, a functional explanation of x is something very different from the description of x's function for in the former we are not trying to identify functions but rather to set forth the effects of a phenomenon in order to give an account of its presence. A functional explanation "could explain why a certain event occurred, why a particular thing has a certain property, why something behaves in a certain manner, and so on, without restriction" (Cohen, 1980, pp. 255–256).

Having thus presented the difference between a description of functions and a functional explanation, it must be clear that functional explanation is produced when we give an account of x as providing the effects generated by x. From here that functional explanation is known also as an explanation by the consequences or by the effects of something. It is worth noting that a functional explanation is not the exact reciprocal of a causal explanation, that is, if one gives an account of e by the fact that it induces f, this does not mean that the form of the functional explanation is: e happened because f happened.

> If that were its form, functional explanation would be the exact opposite of ordinary causal explanation, and it would have the fatal defect that it represented a later occurrence as explaining an earlier one. Nor may we say that the form of the explanation is '*e* occurred because it caused *f*'. Similar constraints on explanation and time order rule that candidate out: by the time *e* caused *f*, *e* has occurred, so that the fact that it caused *f* could not explain its occurrence. The only remaining candidate ... is: *e* occurred because it would cause *f*, or less tersely but more properly: *e* occurred because the situation was such that an event like *e* would cause an event like *f* (Cohen, 1983, p. 115).

Notwithstanding Cohen's comment transcribed above according to which functional explanation gives an answer to the question, why? (in such a way that he can consider that explanations of this form constitute a special class of causal explanations, in contrast with other authors, such as Wood, for example), it is legitimate to doubt whether it is so. *Prima facie*, one has the feeling that these so-called explanations answer rather to the question, what for? and, from the beginning, it seems impossible to sustain the identity between the questions, why? and what for? To clarify this point it is convenient to analyze the various meanings with which the term *function* is used in scientific discourse. The ambiguities of the term are immediately shown. The term is currently used to signify: (a) relations of dependence or interdependence between two or more variables; (b) processes or operations realized by a certain entity; (c) recognized use of something, or the expected effect of an action; (d) consequence of a thing or of a determinate activity for the system to which they belong, or for other components of the system; (e) contribution of an element to the preservation of a certain feature or condition of the system to which it belongs (cf. Nagel, 1974).

In all of the above meanings, we may assume that the identification of the function of x permits us to know something about the aim of its presence, but it says nothing about the conditions in which such a presence is possible.

> Functional analysis is a modification of teleological explanation, that is, of explanation not in reference to causes which 'will produce' the event in question but in reference to ends that determine its course (Hempel, 1965).

The matter rests on whether this information can be characterized, strictly, as explanatory. Of course it is not a matter of proposing a debate on mere terminology. Someone might choose to call 'explanation of x' all the knowledge that can be accumulated about the nature of x, its ends, etc. Nevertheless, that knowledge will not allow us to take one step towards finding the answer to the question, why was x possible?, why is x there? It is also clear that *explanation* can only be said of a statement answering these questions ... what is relevant, in any case is to distinguish the specific character of what is asked in each of the types of research.

It is worthwhile to examine Cohen's favorite example of a simple functional explanation:

> In some industries there is, over a period of time, a marked increase in the median size of the producing units: small workshops grow into, or are replaced by, large factories. The increased scale reduces the costs of producing a given volume of output. It generates economies of scale. If we find that scale grows just when growth in scale would have that effect, and not otherwise, then it is a plausible explanatory hypothesis that the scale grows because the growth brings economies (Cohen, 1980, p. 280).

There is no doubt that sometimes, yet not always, the augmentation of scale generates savings, just as it may produce other consequences: a greater organic composition of capital, a relative decrease of the labour force, control of a larger share of the market, and, maybe, even undesirable consequences, such as a reduction in the profit rate, a greater time for capital rotation, overproduction problems, etc. Let us assume that there is no major difficulty with the fact that sometimes e (increase of scale) happens without f (savings) being produced, and that an explanation can be construed in such a way that it gives an account of the eventual non-occurrence of f; suppose also that there is no difficulty to extend the explanation of e by f to include the favorable consequences; suppose, finally, if its acceptance is more difficult, that the explanation can disregard the undesirable consequences in some manner. It rests, though, on the problem of knowing, if the explanandum is e, whether the explanans must determine why e happened 'at a certain stage' and not at another or, what amounts to the same, what were the circumstances by virtue of which e was possible at a given moment. This question has no answer in the functional explanation no matter how large the explanans is made, that is, the set of propositions which refer to *f*.

Cohen notices such explanatory insufficiency but is not inclined to draw the inevitable consequences. Thus, for example, he points out that

> it is a plausible generalization that a society develops and/or sustains a religion when a religion is necessary for (or would contribute to) its stability. The religion of a society might, then, be explained in terms of this feature of the society: it requires a religion to be viable ... Now suppose a society requires a religion for stability, and has a religion fulfilling that need. It does not follow that its need for religion explain its having one. The society may indeed require a religion, but it is a further question whether it has one because it requires one. It may have one not at all because it needs one, but for other reasons (Cohen, 1980, p. 281).

It is always *another question* and there are always *other reasons* for allowing the answer to the question, why *e*?, that is, *e* and *f* can never wholly be explained. Hence, the claim that religion could be explained in terms of its role in the viability of society makes no sense. Cohen's own argument concludes, no matter how rich the description of the social role of religion is, we will never be allowed to claim that "there is a religion because it is necessary for social stability" even if, once the causes of social religiousness are known, the functions such a religiosity fulfills can be identified and among others, of course, we actually can discover its stabilizing capacity.

The place given to theoretical elaboration is central in order to decide about the validity of functional explanations. Cohen refers repeatedly to the link that, according to him, exists between functional explanation and theoretical elaborations in biology. He claims that such elaborations are possible developments of functional explanations whose validity, nevertheless, does not depend on the former.

> We can rationally hypothesize functional explanations even when we lack an account which, like Darwin's, shows how the explanations work. A satisfying elaboration provides a fuller explanation and locates the functional fact within a longer story which specifies with greater precision its explanatory role more precisely ... we are frequently certain that *p* explains *q* yet unclear how it explains it (Cohen, 1980, pp. 285–286).

Thus, even if theoretical elaboration is desirable, it does not constitute a necessary element of explanation. Hence, he considers that in the case of biology, the theories of Darwin and Lamarck, for example, are rival versions offering different elaborations of the reasons for which functional explanations are maintained but that they are not alternatives to them.

Cohen's arguments seem to conceive functional explanations as mere empirical generalizations; what else could these so-called explanations be in the absence of a theoretical elaboration? If so, their validity is completely groundless and it is justified to object in the sense that "functional explanations without elaborations of any sort give rise to pseudo-explanations" (McLennan, 1981, p. 59). Indeed, if theoretical elaboration gives an account of the mechanisms through which the *functional fact* is produced, there is no base to sustain an explanation of that type without its corresponding theoretical elaboration. What could be the interest of a functional explanation if one does not have at hand the knowledge about the mechanisms that can justify it in a given case? Elaborations are not, in fact, alternative theories to functional explanation; rather they are the argumentative core where explanatory efficiency lies. Thus, to return to the example of the growth of scale of enterprises, the claim to explain this fact through one of its consequences (it generates savings) leads to pseudo-explanation. The phenomenon is only explained, for example, in the framework of a theory of concentration and centralization of capital. Outside this framework, any of the effects of the phenomenon in question can be enhanced but it will never yield its explanation properly. The reference to the effects of a phenomenon contributes to its comprehension but nothing indicates the conditions of possibility of the phenomenon whose explanation is looked for.

Cohen realizes the *methodological indiscipline* existent in the transit from the thesis "... '*A* is functional for *B*' to '*B* functionally explains *A*' without experiencing any need to justify the step" (Cohen, 1983, p. 120). Now, if, as we try to show here, the ill-named functional explanations constitute, in every case, mere functional descriptions, then the risk is present in every case even if we think we are proposing a pertinent explanation and all the subsequent justifications. Historiographic discourse, nevertheless, is full of formulations in which one believes oneself to be explaining a fact in reference to the effects it is generating and in which biased agents are involved. We must insist: knowledge of the effects of x or of the role of x allows a better comprehension but it does not help to answer the question: why x?

Cohen recalls certain passages of Marx where the latter examines the reduction of the labour day resulting from the Factory Acts and sets forth two reasons that led to its promulgation: the struggle of the working class and the need, perceived by the government, of fixing limits to the capitalists' greed for surplus value, so as to guarantee the stability of the social system. According to Cohen's commentary, "these are cited as separate forces, whose confluence produced the Factory Acts. Marx does not say whether either would have been enough without the other" (Cohen, 1980, p. 294). The example interests him because it would be a phenomenon where q (reduction of the labour day) is partly explained by p (workers' struggle), but also by r (stabilizing role). The relation p–q would be fixed by a classical causal hypothesis, but the link q–r would be settled by a functional hypothesis. That is q is explained in part by the consequences it carries. Without doubting that r effectively helps in the comprehension of q, and also to understand why p could break the resistance to the establishment of q, the truth is that the reduction of the labour day is not explained by its stabilizing role. q happens, indeed, in a system S which requires stabilizing measures. The government operates as an institution that is conscious that q is convenient for its stability, but an event never happens because someone is interested in the benefits derived from it. The functional description does not explain why q was possible even if it points to circumstances that do not hinder but rather facilitate it, namely, the willingness of S to reinforce itself with q. This is what Cohen's general hypothesis seems to underline from the analysis of the following example: substantial changes in the economic structure (not radical transformations) that immediately benefit the subordinate class occur when the subordinate class fights for them and when they increase the stability of the system. If what is defended here is correct, then we may notice up to what point description and comprehension are but necessary moments of the explanation. But we can also note the need to distinguish them and not to subsume one into the other.

What theoretical problem leads Cohen to maintain that the main Marxist explanatory thesis has a functional nature? The problem, crudely stated, is the following: How to reconcile the idea that the (basic) structure explains the superstructure and the claim that, nevertheless, this has effects upon the former; how to reconcile the idea that productive forces explain production relations and the claim that these control the development of those. "If the intended explanations are functional ones, we have consistency between the effect of A on B and the explanation of A by B, and I do not know any other way of rendering historical materialism consistent" (Cohen, 1983, p. 116).

Thus if the central explanations of Marx are functional, it is due, according to Cohen, to the fact that the character of what is explained is determined by its effects on what it explains. That is, according to this version, what Marx is trying to explain has an enormous impact upon what he presents as explanatory: the functional construction makes compatible the causal power of the explained phenomena, and their status becomes secondary in the order of explanation.

It is not yet clear that the problem – whose solution is attempted by Cohen by means of attributing a functional character to Marxian explanations – is legitimate. It is highly probable that Cohen is building a complex discourse to find a solution to a false problem. In fact, the link between productive forces and production relations, or between structure and superstructure, cannot be thought in terms of *explanans* and *explanandum*. If we limit ourselves to certain schematic references (for example the *Contribution to the Critique of Political Economy*) we are caught in an economistic perspective from which we cannot escape through the functional reconstruction of the argument. To believe that the structure *explains* superstructure, or that a certain level of development of the productive forces *explains* the prevailing form of productive relations, is equivalent to placing a set of hypotheses whose fragility does not resist the slightest examination in the center of the theoretical device of historical materialism.

We must admit that in Marx himself and in the ensuing development of historical materialism there is, in fact, a strong tendency to construct a social theory of economistic nature. A good deal of the theoretical debates within Marxism have been produced within this tradition. Cohen's effort is oriented towards a reconstruction of historical materialism that reinforces in a notorious manner such an economistic nucleus, based on a distorted reading of well-known formulas that appear in the prologue of the text mentioned above. In it, Marx tells us that "the general result that, once achieved, served as conducting line" in his studies of political economy, can be briefly stated in a handful of propositions:

> in the social production of their life men enter into definite relations ... of production which correspond to a definite stage of

> development of their material productive forces. The sum total of these relations constitutes the economic structure of society, the real basis, on which arises a legal and political superstructure and to which definite forms of social consciousness correspond. The mode of production of material life determines the process of social, political and spiritual life in general. It is not the consciousness of man which determines his being, but rather, its social being which determines his consciousness.

A simple 'general result', a mere 'conducting line' of the study, is transformed by Cohen into the dorsal spine of historical materialism. Formulas in which vague mechanisms of correspondence, conditioning or determination are stated become causal explanations of a peculiar type. "In each case Marx distinguishes two items, the second of which asserts to be in some way explanatory of the first" (Cohen, 1980, p. 278). Economism consists, precisely, in the belief that the economic structure explains the judicial and political superstructure, where the prologue barely states, through a useless and misleading architectural metaphor, the articulation between structure and superstructure, Cohen finds a thesis that explains one by the other. We must insist, then, on the fact that it is impossible to produce any explanation of judicial and political institutionality in a society by building an *explanans* that turns to the features of the economic structure. Worried by the need to reconcile what he takes to be the principal theses of historical materialism (1) the level of productive capacity explains the nature of the economic structure, 2) economic structure explains the nature of superstructure, 3) economic structure promotes the development of productive forces, 4) superstructure stabilizes economic structure, worried that is by the need to reconcile the first two theses with the latter two, Cohen passes over their dubious validity. We must not be surprised, then, that if, in a *personal interlude*, Cohen confesses "in the five years which have passed since *Karl Marx's Theory of History* was submitted to the publisher I have come to think more critically about historical materialism. I have not rejected it, but I have developed doubts about it" (Cohen, 1983, p. 125). In fact, if the theoretical nucleus of historical materialism were constituted by these four theses, as Cohen believes, then it should be rejected.

For, indeed, the arguments constructed upon these stated theses are completely unsupportable. Thus, for example, Cohen writes:

> Now since (1) the level of development of productive power determines what relations (that is, what sort of economic structure) would advance productive power, and (2) relations which advance productive power obtain because they advance productive power, it follows that (3) the level of development of productive power explains the nature of the economic structure (Cohen, 1983, p. 118).

Nevertheless, it is obvious that the second premise is false; no economic structure is established by the simple fact that it promotes productive capacity. And so, it follows that the conclusion is also false as even a superficial examination of historical experience can show. Now, if the four theses enumerated above are false, what is at stake is not historical materialism (although its economistic tendency is indeed at stake ever since Marx), but, precisely, functional explanation since it is introduced upon these theses and to reconcile them. For this reason, one can easily multiply references to formulations by Cohen that lack any historical sensibility and that are posed without being worried by the fact that they have a simple empirical refutation. A final example:

> the prevailing production relations prevail because they are relations which advance the development of the productive forces. The existing level of productive power determines what relations of production would raise its level, and relations of that type consequently obtain (Cohen, 1983, p. 117).

Both facts are false: sometimes productive relations obstruct the development of productive capacity, and the establishment of new social relations is never the direct consequence of requirements derived from the growth of productive capacity

The argument elaborated by Wood to defend the thesis that the explanations proposed by historical materialism are typically teleological offers the same kind of difficulties. Explanations of this type are produced to give an account of the behavior of systems oriented towards the achievement of a certain end; these exhibit a persistent tendency to maintain a determined state or to change in a specific direction.

> We explain some element or aspect of such a system teleologically when we show how it manifests or contributes to the persistent tendencies which characterize the system, and provide reasons for thinking that this element or aspect exists

> because it manifests or contributes to those tendencies (Wood, p. 104).

Here, again, the misuse of the term *explain* leads to a surprising analytic poverty. The description of the functions of x, the way in which x is part of a totality, and the role that it plays in it are confused with the explanation of x. Marx wants, according to Wood, "to explain the character of a society's legal system, politics and moral or religious beliefs by showing how they serve to sanction its social relations" (Wood, p. 105). Yet it should be clear that even if it were true that the set of laws and beliefs is sufficiently identified when its sanctioning role is recognized, this does not allow any further step into the explanation of why such laws and beliefs exist instead of others unless one presupposes – but it is going to be an inadmissible supposition – that such a determined set and only that determined set has such a quality.

If it were true – as Wood believes – that "Marx proposes to explain the social relations prevailing in a community by showing how they manifest or contribute to its tendency of making efficient use of its productive powers (Wood, p. 105) and, furthermore, that Marx would consider this tendency as a "sufficiently basic and pervasive feature of societies to account for the existence of the phenomena which manifest it" (Wood, p. 105), we *would* be facing the unbelievable case in which a single vague idea would have an unlimited explanatory efficiency. We have an example of the sterility of teleological explanations in the fact that Wood is able to write: "revolutions occur and class struggles have the outcome they do, in order that the productive powers of society may continue to expand, and the social relations of production may be suited to them" (Wood, p. 105).

Maybe there is no need to note that in another context, the word *tendency* is used to make reference to a need less strong than absolute, that is, the word serves to think the way in which a connection, postulated as rigorously necessary in an ideal model in which abstraction has been made of many variables, works in reality. Thus, in that context *tendency* pertains to a probable connection. Nevertheless, the term is used here in the worst speculative tradition in which it became almost exclusively derogative to characterize a conception as *teleological*. Here, in fact, *tendency* claims to refer to an omnipresent feature of the system, no matter what the conditions in which it evolves. Once a certain tendency is located, we would believe ourselves able to be allowed to claim beforehand the direction in which such a system will move in the future. For this reason, something as senseless as the claim that class struggle has a predetermined outcome can be sustained, that is, an outcome imposed by the tendency proper to the expansion of productive forces. "Marx's theory rests [Wood assumes] on the idea that there is a general historical tendency for productive powers to be used efficiently, and to expand" (Wood, p. 109). Regardless of whether Marx rests his theoretical venture upon this idea or not, it seems obvious that it corresponds to a simple evolutionist view from which nothing in the historical process can be explained. The thesis that there are revolutions "in order that the productive powers of society may continue to expand, and the social relations of production may be suited to them" is, by itself, an overwhelming example of the damages that is generated from a theoretical discourse by the substitution of the question, why? for the question, what for?

If functional explanations, such as they are defined by Cohen, are, in the best case, a pertinent device for the description and comprehension of the facts to which the *explanandum* alludes, teleological explanations, as defined by Wood, lead to a vacuum of knowledge. Both authors, for the rest, displace the epistemological problems of historical materialism in a direction that can only generate the weakest side of Marxian discourse and stimulate the economistic perspective, by itself so enhanced in the tradition of thought inaugurated by that discourse.

NOTE

[To our regret, the author's manuscript is incomplete and does not have his list of references. The exact references with titles are therefore unavailable for these well known cited authors – Ed.]

CORINA YTURBE

THE HISTORY OF SCIENCE: INTERNAL OR EXTERNAL?

1. Contemporary historians and philosophers of science have shown an even greater reluctance to accept a dichotomy between 'internal' and 'external' factors that intervene in the development of scientific knowledge. In the recent Anglo-Saxon tradition, we find two fundamental positions with respect to the internal-history/external-history problem. One of these consists in defending an 'internalist' position with regard to the history of science, or, at least, in supposing that internal factors have a greater relevance for science than external ones. On the other hand, there is an increasingly more serious interest in constructing a 'new history of science', based on the complementarity of the internalist approach and the externalist approach, and on the utilization of methods drawn from other social disciplines.

In both the historiography of science and the philosophy of science of recent years, there is a tendency to surmount the internalism/externalism debate, and to seek an intermediate point between the two approaches. However, as Mikulinsky[1] points out, since both internalism and externalism are one-sided schemes, neither can serve as a solid theoretical base for the history of science, and, for that reason – as we shall see later – their mere synthesis is totally inadequate for the development of a true alternative.

In this paper I will try to show in what sense the attempts to construct a purely internal or a purely external history of science have proven impossible. I will, moreover, discuss some of the difficulties faced by positions that defend a combination of internalism and externalism. Finally, I will attempt to reveal the extent to which the conception of scientific knowledge drawn from Marxism, especially the Althusserian current, opens a fruitful theoretical path for the development of a methodology for the history of the sciences. An analysis of the epistemological theses that underly these conceptions of the history of science will reveal their theoretical effectiveness as well as their ideological commitments, which constitute true obstacles to the construction of such a history.

S. Ramirez and R. S. Cohen (eds.), Mexican Studies in the History and Philosophy of Science, 71–87.

2. The dominant current in the history of science up to the thirties (and perhaps even today[2]) is undoubtedly the internalist. This approach fundamentally consists of the claim that science develops exclusively by its own immanent laws. Internal explanations are based on the study of experiments, observations, and theories, without the intervention of external factors – whether economic, political or social – in the structure or development of scientific knowledge. There are two principal versions of internalism: one of an empiricist bent, and the other developed by those whom we may call 'historians of ideas' or 'intellectual historians'.

According to positivism, historiographic study is concerned with published works or finished versions, which present only the final scientific propositions, together with the evidence for those propositions. What is omitted is a discussion of the way scientists arrived at those propositions – that is, the manner in which concepts and hypotheses were modified – as well as any extra-scientific factor. Moreover, it is held that the logical and methodological structure of science is what enable us to explain its 'internal rationality'; that is, only 'intellectual' factors are able to influence the character of science. In other words, the history of science is conceived as a rational reconstruction, or as a history of mere ideas – *rationality* is identified with *ideas*. Furthermore, knowledge is not only developed according to its own internal logic, but in a cumulative manner: scientific activity is presented as the gradual and continuous development of laws, based on rigorously controlled observations; and the idea of revolutions or ruptures in science is rejected. History is constructed as a mere chronicle, as a presentation of 'facts'; and any interpretation is avoided.

In the face of this 'history of facts', the other current, the idealist tendency in historiography, also conceives of the history of science as a history of ideas – that is, as a history in which only intellectual factors intervene – but, in this case, we attempt to give an interpretation of such a historical development. Intellectual historians (as they are called) adopt this approach, maintaining that the objects of a history of science are intellectual structures. The theoretical system of each period is developed in accordance with its own internal logic, and in order to understand it, the historian has to try to 'rethink the thoughts of the scientist', to learn to think in terms of the scientist's own rules of argumentation, and to use the scientist's own criteria of truth.

As with the historiography of this tendency in general, there are two important obstacles to an effective internal explanation of the history of science: first, the subjectivism which arises from this psychological approach – which takes the capacity of the historian to 'rethink thoughts' as a basis of explanation – and, secondly, the idealism present in the Hegelian idea that history is fundamentally the history of thought, seen as a sequence of intellectual structures developed in accordance with internal rational criteria.

A. Koyré adopts this idealist position, wherein intellectual factors are taken to be the only determinants in the historical development of science. According to his position, we do not try to think like those in the past; rather, we try to clarify how scientists thought in the past, by making explicit the rules of argumentation and criteria of truth. While the idealist (subjectivist) historians attempt to recapture the past by 're-living it', Koyré seeks to clarify it by utilizing present-day categories. His work is fundamental, inasmuch as it provides the theoretical framework for the development of a professional and rigorous new history of science. Further, although he explicitly maintains idealist theses regarding the explanation of the development of scientific knowledge, such as the total autonomy of such development with respect to external factors, his explanatory program gives a place to a theory of science that affirms the existence of discontinuity in the development of scientific knowledge. For Koyré, there are revolutions in rationality, and not only in theories, and he recognizes the essential role of extra-scientific ideas – such as metaphysical conceptions, religious beliefs, and magic. In this way, Koyré offers us an alternative both to positivism and to extreme externalism.[3]

Nevertheless, the tendency in the Anglo-Saxon world to continue taking Koyré's work as a model to imitate in the history of science has been an important hindrance in the development of a new history of science which would be a true alternative to that connected with positivist philosophy.

3. The internalist approach in the history of science maintains that, in virtue of the autonomous character of science, factors which we may call external do not exercise any decisive influence on the structure of science – at most, external factors merely serve as the backdrop against which scientific events take place. The externalist approach, on the other hand, is characterized by the attempt to locate scientific activity in an economic and socio-political context that would enable us to understand both the development of science and its social function. The emphasis

of externalism, then, is on the economic, political, and social factors that influence the development of scientific discourse.[4]

There are various reasons for the change in approach in the history of science. Thus, for example, it is more and more difficult to deny the influence of factors originating in economic, political, military, industrial and ideological sectors, as much on the production as on the utilization of scientific knowledge. While traditional history of science is characterized by its "massive disconnection from important problems concerning scientific development and its consequences",[5] it is increasingly more evident that factors considered 'external', such as scientific policies, or the function and place of scientific knowledge in society, are aspects that condition at its interior the development of such knowledge.

A consensus does not exist concerning how we should understand this type of history of science, characterized as *external*, in contrast with the *internal* approach. One consequence of this lack of a common program, with solid theoretical bases, is the confusion of this approach with other positions that are unacceptable, and this tends to lead to either explicit or implicit rejection of this approach.[6] If the fundamental failure of the internalist conception is to be found in the opposition that it establishes between thought – the context in which it considers science to develop – and socio-historical aspects, then the theoretical weakness of externalism (in its extreme version) is that it establishes a mechanistic type of causal relation between the historical conditions of science and its internal structure, reducing scientific activity to a mere phenomenal manifestation of the socio-economic relations of a particular society.

In reality, what leads to the rejection of the external history of science is as much the vagueness in the presentation of this analytical program as its translation into simplistic Marxist theses which end up denying the relative autonomy of scientific processes.

4. Even though a great many historians of science appear to be uninterested in resolving the basic epistemological problems present in their discipline, it is becoming more and more evident that there is a fundamental connection between the history of science and the philosophy of science. The different ways of constructing a history of science are – most of the time – based on explicit or implicit commitments to philosophical theses concerning history, science, and their mutual relation.

One of the philosophical theses in favor of the contraposition between the internalist approach and the externalist approach is the so-called doctrine of the two contexts, developed by positivism. This conception constitutes the dominant framework in which the philosophy of science has been developed, and consists in drawing a radical distinction between the context of *discovery* and the context of *validation.*

The context of 'discovery' includes a variety of elements: from the individual psychological processes that lead to the 'discovery' of knowledge, to sociological, economic, ideological, and political aspects that condition the production of knowledge. All of these elements are characterized as non-logical or non-rational. Since it is difficult, if not impossible, to establish laws or generalizations concerning all these subjective aspects, the context of discovery is considered somewhat obscure, vague and, up to point, not theorizable. Thus, the context of validation – the analysis of the logic of science – became the only aspect relevant to the history of science. The consideration of non-rational, external factors, according to this position, not only fails to increase our understanding of scientific development, but even obscures the fundamental question of the rational validation of scientific arguments.

From the positivist perspective, then, one began from a static view of scientific knowledge, ignoring any socio-historical element, with the consequent disconnection between history and philosophy of science. While the philosopher of science would occupy himself only with the validation of scientific knowledge – the only legitimate problem – the whole of the context of discovery was left in the hands of the historian who, in the best of cases, would have to resort to other social disciplines in order to give an account of it. In general, the result was that the dominant way of doing history of science consisted in reviewing the elements present in a cross-section of scientific knowledge, assuming that external factors have no influence in the rational structure of science.

Even though important problems appeared within the positivist philosophy of science, which made it take new directions, other questions related to scientific activity in general, its production and utilization, remained outside of the field of study of the historiography of science that continued being positivistic. Disciplines other than the philosophy or history of science – such as sociology, economics, etc. – began to concern themselves with the socio-historical and economic aspects of science, taking into account the fact that any scientific discourse is written within a particular social reality, in which it fulfills a specific social function, and by which it is determined.

In this sense it can be claimed that, in concealing relevant factors, the doctrine of the two contexts fulfills an ideological function both within the discourse of philosophy of science and with respect to the internalist history of the sciences: "the independence of the framework of validation acts in fact, and 'by right' besides, to delimit the field of professional action in the philosophy of science . . . , attributing, then, irrelevance to a set of factors . . . that determine scientific production".[7]

Nevertheless, problems arising from within the context of validation, as well as problems related to the conditions that make possible and orient the production of scientific knowledge, led to the serious questioning or abandoning of this conceptual framework of the traditional philosophy of science.

4.1. Among the elements that have influenced from outside the criticism of the two contexts, the most important are those related to the production of knowledge.[8]

Beginning with the Second World War, scientific development has revealed the relevance of the conditions of production for the study of its validation, and, consequently, the impossibility of examining the context of validation as if it were self-sufficient.

In other words, the production of scientific discourses appeared related to other interests – such as political, ideological, military, and economic – distinct from so-called 'pure' scientific interests. It cannot be denied that, in the advanced capitalist countries, science had to convert itself into a directly productive force, a tendency that has obvious consequences in the production and utilization of scientific knowledge. The socio-economic structure of those countries demanded certain types of knowledge and technologies, which had the necessary consequence that both the lines of investigation and their results tended to satisfy those demands. It therefore becomes very difficult to continue thinking of scientific knowledge as determined exclusively by its gaps in knowledge or by the progressive refinement of its explanatory hypotheses. The planning, financing, and direction of scientific research, among other things, show that the effect of scientific discourses cannot be reduced simply to their function as expressions of knowledge. If, "*scientific policies, whether independent or not, everywhere determine the course of research, whether applied or not*",[9] the inescapable confirmation of this fact would occasion an important change in the manner of conducting studies in the philosophy of science, and would have a striking effect on the historiography of the sciences.

On the other hand, the doctrine of the two contexts not only would fulfill the ideological function of concealing the conditioning of the production of scientific knowledge by requirements that arise from economic activity, the political system, the military apparatus, etc.; but, moreover, the treatment of science as something determined only by intellectual factors is connected with the ideological function of the dominant political discourse which also has to conceal the relations between knowledge and power.

4.2. It was not just the unavoidable presence of the problematic relating to the production and application of scientific knowledge which influenced the split in the traditional conceptual framework – positivist or idealist – upon which philosophy and history of science had been developed. The very task which traditional philosophy of science had proposed, namely, the rational reconstruction of scientific systems, is faced with serious problems that arise with respect to the context of validation itself.

In general, traditional philosophy of science departs from the fact that change is rational – or, at least, rationally reconstructible – and proposes different ways by means of which it seeks to explain not how scientific activity actually proceeds, but rather how it would have to proceed if science is to continue being a rational activity. Its field is, then, the rational reconstruction of the series of products of the rational activity that is science. To place confidence in the explanation of the course of the history of the sciences that essentially refers to the action of social or historical factors would be nothing but a show of irrationalism. The process of production of a theory would be, so it is said, the task of a fundamentally psychological analysis. Social problems of any sort are impossible to present.

In this paper I will not discuss to what extent the different rational reconstructions offered to this day are acceptable or not. I only want to point out that, independently of the omission of socio-historical aspects, the concept of validation itself – the core of the conception of the two contexts – is quite clearly insufficient, "it is not only necessary to take into account the elements of the production of knowledge (scientific policies above all) in order to consider validation in an adequate way, but further, it is at the very heart of the 'classical' epistemological proposals that we find a profound deficiency."[10]

We cannot ignore the fact that, in its day, the philosophy of science with its neo-positivist roots constituted a notable stimulation for meta-scientific studies, highlighting the methodological importance of the recourse to logic and to experience. But this philosophical position leads, by way of the philosophical extrapolation from formalism and empiricism, to a vision of science at odds with its reality, which is what I am interested in emphasizing for its importance for the history of science.

We cannot speak of a new way of doing philosophy of science, and of the approach of philosophy to the history of science, without mentioning Kuhn. The great merit of Kuhn consists in his having rescued the history of the sciences as a working tool for the explanation of scientific activity. Starting from his perspective, and, in particular, from his work, *The Structure of Scientific Revolutions*, the development of scientific knowledge begins to be treated as non-cumulative, and as related with history more generally.

While the radical distinction between the context of validation and that of discovery, and the resulting dichotomy between internalism and externalism, continues in the history of science, Kuhn believes that the effort to unite these two approaches is, perhaps, the greatest challenge that present historians of science face.[11] At the risk of being regarded as 'irrational', Kuhn centers his investigation of the development of knowledge on the analysis of 'scientific communities' in whose subjective (or intersubjective) criteria resides, in the last instance, the scientificity of a particular theoretical discourse.[12] While in the early stage of the development of a new field of scientific activity, or in the inauguration of a new science, "social needs and values are a major determinant of the problems on which its practitioners concentrate", in the mature sciences the members of the scientific community are the "exclusive audience for, and judges of" the products of their activity. The scientific community is, then, the subject of scientific knowledge, the subject that determines the problems to be dealt with, the lines of research to follow, and, above all, the scientificity of its activity. Thus, scientific communities are found *isolated* from society and its effects. In its mature stage, economic subsidies, institutions, military demands, and ideological conceptions no longer intervene in the structure of a body of scientific knowledge.

The work of Kuhn, in virtue of its approach to history, in effect puts in relief problems that the philosophy of science cannot ignore, and points toward a new way of analyzing the sciences, which fundamentally questions the stereotypes of positivist origin. Nevertheless, it is precisely the psychologism and sociologism present in his work which impede the consideration of the development of scientific knowledge within the framework of the social conditions of its production.

5. With the recognition of the importance of the relation between philosophy of science and the history of the sciences, it has been much insisted that we will not be able to obtain from a philosophy of science based on positivism, inductivism, or conventionalism, a satisfactory response to the question concerning the relations between factors *internal* to a scientific system and the socio-historical, that is *external*, factors that intervene in the development of scientific knowledge.[13]

In contrast to the philosophical analyses of the structure of science that arise from those tendencies, the historical character of science, the non-cumulative development of knowledge, and, in some cases, the social character of scientific activity and, consequently, the influence of socio-historical aspects on its development, have been especially emphasized. But, in spite of this shift in emphasis, a satisfactory program for the study of the history of science still has not been produced.

The search for new explanatory programs is characterized above all by the attempt to reconcile the internalist and externalist approaches. But, in view of the fact that both approaches are committed to theses concerning the nature of science that are not only incompatible, but are unsustainable, their union *without important changes* in their philosophical presuppositions cannot result in a viable program.

Among the principal problems with which these new projects are faced, we may mention both the difficulty in establishing a clear boundary between internal and external factors, and the diverse epistemological and methodological questions that arise from the utilization of methods of analysis originating in other disciplines without definition of the theoretical object of the history of science.

Thus, for example, some of these projects end up considering the *external* factors as *historical*, but the internal structure of science continues to maintain its ahistoricity. To speak of 'the sociology of science', 'the theory of roles', 'the social history of science', without making clear what would be the place of these disciplines in a historiography of the sciences, does not appear to be the most adequate solution to the internalism/externalism dichotomy. We could say of these attempts what Mikulinsky says of the externalist approach, that they would imply that the historical and social conditions appear as purely external in relation to the internal structure of science. The great problem would continue to be the impossibility of clearly stating the influence of these factors on the internal character of science.

6. The work of the Althusserians owes its importance to their renovation of the discussion of questions central to Marxism, as is the case with the problem of the production of scientific knowledge. Many of the formulations of their proposals are insufficient, and a great part of their value lies not so much in offering a solution as in indicating the place and direction in which one ought to present the problem in question.

Departing from the thesis of Bachelard that a history of the sciences, because of its very object of study, implies an epistemology, one of the principal tasks consists in eliminating the obstacles that stand in the way of the construction of the concept of the history of the sciences. The attack is directed against three philosophical affirmations which have contributed to slowing the development of a historiography of the sciences: that science is unitary, that its evolution is continuous and uniform, and, finally, the interpretations concerning the 'essence' of science, which are present in the definition of the 'object' of the history of the sciences conceived as a history of methods and results (as in positivism, inductivism, and conventionalism). Insofar as the object of which the history is going to be made is not defined, the history of the sciences disappears as such, and is replaced by a diffuse chapter of the history of ideas.[14]

The aim of the Althusserian position is to consider the problem of the sciences as a problem of ways of functioning. It rejects the existence of Science in general, and affirms that what exists is a process of knowledge, developed as a function of concrete problems that are presented at particular moments in a particular society. In other words, the sciences are thought of as systems with relevant effects, and hence, with specific differences: by 'science' – that is, scientific practices – we should understand a particular form of social production whose specific product is knowledge.

There are no processes of knowledge production in a 'pure state', there are only defined ways of producing knowledge, conjoined with other production processes (scientific or non-scientific) within a concrete scientific practice, in turn conjoined with other social practices in a concrete society. The objective of a history of the sciences would be, then, the explanation of the process of production of concrete knowledge. The answer to the question of what defines a process as an effective production of knowledge – that is, as science – can be drawn only from an analysis of its concrete conditions of production and efficacy. Through their development, then, the sciences construct themselves at the same time that they construct the results with which they modify the concrete social practice in which they lie.

In order to be able to think about the specific nature of this knowledge-producing practice, it is necessary, therefore, to consider the relation that exists between this practice and other practices, and, at the same time, to consider the way in which the sciences are conjoined with social reality. That is, it is necessary to consider the way in which different scientific practices intervene in social practice; for example, we must give an account of the relative independence of scientific practice from ideological practice, as well as from the political-economic function of the science, once we consider the sciences as social practices realized in institutions and intervening materially in the process of economic production.

Another reason for speaking of 'scientific practices' is the fact of the historicity of scientific activities: the knowledge produced by scientific discourses is established through a process which is socially determined. The foundation of a science does not appear in a progressive manner, but is the result of a theoretical revolution, which, on the base of new forms of social practice, produces a totally different problematic that is always open to further development.

The category of 'epistemological rupture' permits us to think about this change of problematic: knowledge is not simply cumulative, nor does it progressively approach the 'truth'; rather, there are qualitative changes, and there are breaks or ruptures which result, at a particular moment, in a completely different theoretical world. Once the science is constituted, its development might very well be progressive, up to the point at which there is again another rupture, and so on successively. The break, as Althusser has pointed out, is the "beginning of a process that has no end ... ", "the break has to be considered immediately, ... not only as an event, but also as an ... internally contradictory ... tendential process".[15]

It is necessary to insist on the *relative autonomy* of scientific practice; this contrasts with the view that results from an assimilation that tends to be made of the Marxist position with the (extreme) externalist affirmation, that the rise and development of science are determined only by the necessities of production and of the development of the productive forces of society. Science is a determinant and singular practice, in virtue of the nature of the object to which it is applied, of the working instruments which are uniquely its own, and of the social relations in the context of which it produces. Obviously, the manner in which these different elements combine is specific to each science. For that reason, scientific practice possesses a relative autonomy, a particular development, according to its specific structure, and cannot be reduced to any other practice.

The autonomy of science is *relative*, since it depends on the position that this practice occupies in the conjoined system of all practices; that is, scientific practice is not without its necessary relations with other practices. "If the sciences, production, and the arts, etc., have a history, that is, a precisely autonomous process of transformation, this is precisely because of the *fact* of their reciprocal determination."[16] To explain science in this way presupposes locating its relative autonomy and its connectedness in the heart of social practice. According to this conception, the meaning and function of scientific practice are not accidents, nor are they owed to an individual 'will', but are determined by the fact that science is developed in the framework of a given social structure.

On the other hand, knowledge – a specific effect of scientific practice – is expressed in the form of a theoretical discourse, which, if it is to be considered scientific, has the character of a system. That is, a scientific discourse controls, according to certain rules, the decisive elements which intervene in scientific production, revealing the way in which such elements are connected and hierarchized within the scientific system. The rigor and efficacy of concepts, hypotheses, methods, etc., depend on their place and function in the heart of the system which is the scientific discourse. Each scientific theory contains, then, its own, historically determined, criteria of scientificity.

This position acknowledges the historicity not only of knowledge, but of epistemological criteria as well. Thus, there is no rationality-in-general, and the criteria of validation, besides being immanent to each theoretical field, are immanent to each moment of the history of the formation of such discourses. Now this is quite different from maintaining a relativism concerning science; it is simply to recognize that scientific constructions can never be postulated as definitive. Scientific work is always conducted in a determinant situation, and the knowledge produced is always exposed to refutation and revision in important respects, without ever arriving at final and absolute knowledge.

Neither can there be, then, a History of Science, since this would presuppose that there exists a general scientific process, applicable to all scientific practices. If science is not unitary, if its evolution is not continuous and uniform, if each science possesses a relative autonomy, then what we can have are histories of the sciences which describe the real processes of the production of knowledge.

7. How, then, are we to think about the relation between internal and external factors, from this point of view?

In the first place, if the task of historians is to give an account of the evolution, movement, transformation, or change of its object, it is evident that a history of the sciences cannot give an account of its object if it does not take into account the so-called 'external' factors, to which externalism refers. Now if science is a mode of social practice, that is, if theory is itself a practice, and thus necessarily related to other practices, these factors should no longer be considered as 'external'; rather, they should be seen as forming a part of the very structure of science. That is, if we understand science as a social phenomenon, then scientific development will be an inseparable component of social development. On the other hand, as we have already noted, we should not ignore the relative autonomy of science, understood as a system of scientific knowledge.

A history of the sciences would have to give an account both of the relative autonomy of science (of its 'internal' structure) and of the conditions that intervene in its production and social function. In a very general way, we can regard the components of the object of the history of the sciences as the following:

(a) The *conceptual apparatus* of a particular scientific discourse. The history of the sciences has to give an account of the real conditions of production of its concepts, insofar as

> a science is not born from the definition of an object nor from the discovery of an object, nor from the imposition of a method – it is born from the constitution of a body of concepts with their corresponding rules of production. Thus, the evolution

> of a science consists of the formation of its concepts and theories.[17]

The problem of the history of the sciences is not, then, to give an account of the acquisition of knowledge viewed as a set of results, but to give an account of the movement of the concepts and theories, which in each science can have types of constitution or of formation which are impossible to assimilate under a single model; that is, they are essentially *historical*. This development of concepts, methods, etc. is fundamentally the result of laws internal to the structure of the scientific practice.

(b) Among the complex set of particular processes that intervene in the production of knowledge, we should mention the different *problem fields* from which the different theoretical discourses are gradually formed. A history of the sciences should explain the transformation and evolution of these problem fields, which define the questions, problems and responses of a particular scientific discourse, identifying the different theoretical problematics out of which the problems of the science in question are gradually constituted. Here, as with the conceptual apparatus, there are improvements, corrections, critical refutations, and cumulative processes.

(c) A history of the sciences also has to give an account of the *matrix from which the different problem fields are constituted.* Although the production of concepts and problematics is organized and regulated by a system of internal norms, the formation of problem fields frequently arises from ideological intuitions or in response to a political or social situation. That is, the development of a problem field does not depend only on the inability of that field to respond to certain problems, nor only on the internal logic of the science – it depends on factors originating from the 'exterior' of the scientific practice as well. In other words,

> it is necessary to study the relation between the theoretical problematics and the given *ideological field* (to use Althusser's term) – that is, the system of ideological formations organically connected to such and such a period in the history of a social formation – since it is this relation that allows us to understand the *contradictions* of a theoretical problematic.[18]

(d) Finally, as we have already pointed out, we cannot give an adequate account of the development of scientific knowledge if we do not see it as conducted in an institutional context, where it fulfills a specific *social function*. A history of the sciences should, therefore, be concerned with the place and function of the sciences in each society, with the relation between scientific practice and the other social practices, with the scientific policies that guide research, and with the institutions (the centers of research and teaching) in which the production and circulation of knowledge takes place. The reality of scientific knowledge is not only in its coherence, but in its *practice*, a practice which depends on the social structure in which it is included. It is for that reason that extra-theoretic determinations of the rhythm of scientific development – such as the social demand for certain kinds of knowledge, and the technical applications and ideological recognition of such knowledge – are important for a history of the sciences.

The general tendency in the historiography of the sciences is to consider the social function of science, as well as some aspects of the matrix from which the problematic is formed, as *external* elements, while the conceptual apparatus and the problem field are treated as *internal*. But we should not think of science as having two independent histories. Science does have specific characteristics in virtue of which its historical development is subject to specific regularities, but scientific knowledge is not developed by its internal logic alone, independently of economic and political conditions; rather, these conditions play a more-or-less determining role in *all* the components of scientific development.

External factors are not only found in the context of discovery, they are present also in the development of the concepts, problems, methods, problem fields, etc. of scientific discourses: that is, external factors are present in the context of validation itself. There are scientific discourses in which ideological conceptions pass on to form part of the body of the science itself, functioning as principles which define its field of study or guide its research; thus, external factors can become internal. This is the case, for example, with biological discourse, in which conceptions of the world and ideological concepts are not simply set aside in the scientific production of concepts.[19]

We can conclude by noting that,

> Economic, social, psychological, logical and other factors are external as long as they remain neutral with respect to the content of science, but they become transformed into internal factors as soon as they begin to interact with the object content of science.[20]

The history of the sciences shows, then, that the historicity of science is not rooted only in extra-theoretic factors, that the internal factors of a specific science are linked to external factors, and that only a concrete analysis of a particular scientific discourse will reveal the specific, and always relative, relation between the external and the internal.

NOTES

[1] S.R. Mikulinsky, 'Internalism–externalism controversy as a phony problem', in E.G. Forbes, *Human Implications of Scientific Advance* (Edinburgh University Press, 1978), p. 88.

[2] T.S. Kuhn, 'The history of science', in *International Encyclopedia of the Social Sciences* **14** (1968), p. 76.

[3] Cf. above all, 'Perspectiva de la historia de las ciencias', in *Estudios de historia del pensamiento científico*, Siglo XXI, Madrid, 1978.

[4] Even though it may be arbitrary to designate chronologically the beginning of a research program, it can be said that the papers presented by the Soviet delegation at the Congress of the History of Science and Technology, which took place in London in 1931, marked a turning-point in the manner of doing history of science. *Science at the Crossroads* (reprint London, Frank Cass, 1971), in effect shows the importance of considering science as a social activity, related with other forms of social practice, and fulfilling a specific function within society.

[5] Cf. M.H. Otero, 'Historia de la ciencia e ideología', in M.H. Otero (ed.), *Ideología y ciencias sociales* (Mexico, UNAM, 1979), p. 216, and Thackray, 'Science: Has its present past a future?', in R. Stuewer (ed.) *Minnesota Studies in the Philosophy of Science, Vol. 5* (University of Minnesota Press, 1970).

[6] Mikulinsky, *op. cit.*, pp. 88–89.

[7] M.H. Otero, 'Producción y validación del conocimiento científico', in *Dianoia* (Mexico, 1977), p. 120.

[8] *Ibid.*, pp. 105–108; we accept, with M.H. Otero, that there is sufficient reason to believe that there is a relative independence between the subjective factors (i.e., psychological) and other aspects (such as socio-historical, economic, political, and ideological) that intervene in the production of scientific knowledge.

[9] *Ibid.*, p. 110.

[10] *Ibid.*, pp. 115–116.

[11] Kuhn, *op. cit.*, p. 75.

[12] Cf. Kuhn, *The Structure of Scientific Revolutions* (University of Chicago Press, 1962), pp. 64, 200.

[13] Kuhn, 'The history of science', p. 80.

[14] Cf. M. Fichant, *Sobre la historia de las ciencias* (Buenos Aires, Siglo XXI 1975), p. 66.

[15] E. Balibar, 'De Bachelard a Althusser: el concepto de 'corte epistemológico'', in *La filosofía y las revoluciones científicas* (Mexico, Grijalbo, 1977), p. 28.

[16] *Ibid.*, p. 27.

[17] M. Fichant, *op. cit.*, pp. 88–89.

[18] E. Balibar, *op. cit.*, p. 30.
[19] Observe, for example, the role played by vitalism, which in biology represents a *conceptual position* that was influential at certain moments during the construction of this science. Cf. M. Pêcheux, *Sobre la historia de las ciencias, ed. cit.*, p. 39, and G. Canguilhem, *El conocimiento de la vida* (Barcelona, Anagrama, 1976).
[20] Mikulinsky, *op. cit.*, p. 96.

ANA BARAHONA ECHEVERRIA

GENETIC MUTATION: THE DEVELOPMENT OF THE CONCEPT AND ITS EVOLUTIONARY IMPLICATIONS

INTRODUCTION

Nowadays, mutation means a change in the kind, number, or nucleotide sequence of the genetic material (Herskowitz, 1962), a heritable change in a chromosome (Watson, 1977), or a change in the genotype not due to a recombination of genes (Dobzhansky, 1970). Mutation is a term used in a general manner to explain the emergence of a character not inherited from ancestors, but heritable by the offspring.

However, this concept has not always had the same significance. The term 'mutation' was used, for the first time, by the paleontologist Waagen in 1869 to identify the minimum perceptible changes in the temporal series of forms he observed in a species of ammonites. Later, it was reintroduced by Hugo de Vries (1901) to identify the process through which new species came forth, and ever since the concept has had a prominent influence in the development of genetics and the theory of evolution.

Mutation is the source of primary organic diversity (Dobzhansky, 1977). However, while pursuing the development of the concept we may note that the problem was not so much to accept mutation as the generating process of variation, but rather to attempt to define its limits: how, when and where it occurs, and in what sense(s) it can be said that a mutation is a random event.

To say that a mutation is a random event we must first of all, define what is to be understood by 'random'. In this case, the word 'random' often refers to the lack of correspondence between a phenotypical effect and the adaptive direction of evolution. It also indicates that the probability that a gene will mutate is the same for each and all of them.

Closely associated with the concept of mutation we find three possible meanings for 'random': (1) random mutations occur if these describe a uniform distribution, that is, every gene has the same probability of mutation; (2) that the variation range for a character is not limited, that

S. Ramirez and R. S. Cohen (eds.), Mexican Studies in the History and Philosophy of Science, 89–107.

is, all theoretically possible changes can occur; and (3) a mutation is random because it does not have an adaptive direction.

The third meaning is currently the most widely accepted among biologists. Regarding the first two, Benzer (1955) has shown the presence of 'hot points' in the genes, i.e., regions where the probability for mutation is greater, and where the range of phenotypical effects is limited (Muller, 1927b). This means that the phenotypical direction of a mutation can be predicted.[1] In this connection, Ayala and Kiger (1980) have noted that random mutations have three possible meanings: (1) a mutation is accidental if it involves an error in the process of duplication of DNA, (2) mutations are accidental or random only because it is not possible to predict which gene will mutate in which specific cell, and (3) mutations are accidental, or not-directed, because they are not oriented toward adaptation.

As the concept of mutation has been historically linked to the concept of gene, an analysis of the development of the gene is necessary to further a study of the concept of mutation.

If it were established that mutation is the primary source of organic diversity, we would have to recreate the concepts of the gene that have been present through the history of genetics, and to see which are associated with the concept of 'random mutation'.

The purpose of this paper is to describe the historic bond between both concepts and to show how the notion of mutation has evolved, from its introduction by Hugo de Vries, up to the notion of exon as a functional and evolutionary unit. At the same time, we shall try to analyze ideas of mutation and to clarify in what sense mutation can be considered aleatory.

For this purpose we divide this paper into four parts. This division depends not only on the discoveries made, but above all on the questions raised by experimentation and by the different hypotheses proposed for understanding the origin of variation and how it is transmitted.

1. MUTATION AND MENDELISM

By the end of the 19th century, the idea of evolution had found a place in the scientific literature. However, Darwin's theory of natural selection as set forth in *The Origin of Species* in 1859, was widely criticized.

Among its critics were experimental biologists such as Hugo de Vries, Bateson and Morgan, who, among others, judged Darwinist theories to be incomplete, unable to explain the origin of species. Two of the most important reasons for their scepticism were the apparent impossibility to solve the problem of variation (which Darwin had ignored) and that of the required assumptions of evolutionary theory.

Darwin's theory of natural selection asserts that more individuals are born than the system can support. Those whose variations are favorable will survive in the struggle for resources, whereby a minimal variation, or difference between organisms, will give some of them a greater possibility of surviving and will allow them to transmit such variations to a larger number of individuals in subsequent generations.

We emphasize that Darwin thought that the existence of variations in nature was necessary for natural selection to take place. Even if he established the existence of two types of variation,[2] he was still unable to provide an answer to the question of their origin, or to explain how these variations are conveyed to ensuing progeny.

Many experimental biologists, late in the 19th and early in the 20th century, doubted that the primary mechanism of evolution was natural selection acting upon continuous variation. Those doubts concerned the absence of a theory of inheritance that could serve as a solid base for the theory of natural selection. At the end of the 19th century such a theory was wanting. At last, in 1900, it seemed that an explanation of how variations were inherited had been found with the rediscovery of Mendel's laws. However, this was not the case. Those who denied natural selection before 1900, supported their views despite the knowledge of Mendelian laws (de Vries, Bateson). On the other hand, those who acknowledged Darwin's theory found few applications for Mendelism within the theory of selection.

At the time, the studies of Hugo de Vries had a most relevant role. He was one of the three re-discoverers of Mendel's laws in 1900, and while endorsing an anti-Darwinist position, he furthered his 'theory of mutation' (de Vries, 1901). This theory claimed that species appeared through a discontinuous process, each new unit, a new step in the process, completely separated from the previous one. New species came forth abruptly, originated from parental species with no preparation whatsoever and with no obvious transitional forms, that is, by leaps (de Vries, 1901).

Against Darwin, de Vries held that his theory of mutation opposes the conception of the selection theory which states that the material substance for the origin of species is provided by individual variations.

According to the theory of mutation, individual variations cannot be the source of new and constant characters since these individual variations fluctuate around a point, and the creation of new species requires the establishment of new forms. In order to explain the emergence of new species, de Vries grounds his view on sudden changes and not on ordinary variation. Accordingly, de Vries defines mutability as the process from which new species are formed, and proposes the term 'mutability' to describe the type of variation capable of producing new species. This forming phenomenon fits all living organisms: all animal and plant features appear through mutation. De Vries assumes that specific differentiation of organisms does not follow a well-defined direction(as in the Darwinist proposition), but rather, that mutations come about in spite of their adaptive value and persevere as long as they are not perilous to, or weakening of individual fertility.

But, how does variability appear? Darwin (1859) said that natural selection was shaped by variation. He thought that it was of a fluctuating nature and that selection chose the forms most fit among those within a spectrum. These variations would be inherited, for otherwise they could not have an evolutionary sense. Unfortunately, Darwin studied complex characters instead of alternative (Mendelian) ones. Complex characters had the tendency to join and to form gradual series, thus bolstering his vision of natural variations as almost imperceptible.

During this period, the concept of mutation was limited to the 'discontinuity' of variation involved in the evolutionary process which made doubtful the notion of evolution as the selection of small variations. This led, on the one hand, to the dismissal of Darwinism as an evolutionary explanation and, on the other, to the possibility of dealing with evolution experimentally through mutations.

The concept of mutation had as much influence in genetics as did Mendelism. The publication in 1901 of the first volume of the *Theory of Mutation* had a great impact on biology because it explained the general problem of the mechanism of evolution, while the Mendelian principles did not yet have universal relevance.

Mutation, explained as the mechanism through which species originate, opposed selection through small variations as proposed by Darwinism, and offered experimentalists an opportunity to repeat or expand the Devriesian experiments into other fields.

The effect of such a concept in theoretical as well as in experimental genetics is clear: the origin of new characteristics through single steps made it possible for genetics and evolution to be treated in an experimental way.

The rediscovery of Mendel's laws introduced the notion of particularized inheritance, that is, a notion of inheritance that claims that differentiating characters are carried by germinal elements. In most theories of inheritance particular units had been proposed as hereditary conveyors: such were Spencer's physiological units, Darwin's gemules, Nagelie's micelar groups, de Vries' pangenes, Weisner's plasomes, Hertwig's idioplasm, Weissmann's biophores, etc. These theories developed from Darwin's theory of pangenesis or from the cytological studies of the nucleus and chromosomes; none were from experimental discoveries.

Johannsen in 1903 replaced the term 'unitary character' by the word 'gene', taken from Darwin's pangene and reintroduced by de Vries. He proposed the term to describe those elements, present in the sexual cells, that would determine the different features of organisms. In 1910 he introduced three basic concepts: 'phenotype', for the attributes of organisms; 'genotype', for the factorial basis; and 'genes' for the factors themselves.

The introduction of the concept 'gene' was an attempt to clear the confusion between the character and its transmission basis, the former being the product of the activity of the latter. Thus, mutation acquired a more precise connotation: heritable changes in the genes. Mutation was thus transformed into the ultimate source of variation. By virtue of the chromosomal theory of inheritance and the theory of the gene, mutation and its role in evolution were honed.

Simultaneously, an important question was posed: what could be the material basis for the Mendelian factors (genes). Sutton (1902) pointed to the similarity between Mendelian segregation and cytological observation of the separation of homologous chromosomes during mitosis and meiosis and suggested that the factors should be the chromosomes or parts of them. An important objection to the idea was raised: there are more characters expressed in adult forms than existing pairs of chromosomes. The solution to this problem came mainly from the work of Morgan and his students at Columbia University during the second decade of the 20th Century.

2. MUTATION AND CHROMOSOME THEORY

At first T. H. Morgan accepted the mutation theory of Hugo de Vries as an evolutionary explanation. His work with *Drosophila* had the initial aim of proving the existence of Devriesian mutations in the animal kingdom (1908). When it became possible to explain his results through Mendelian genetics, Morgan changed his points of view regarding this theory in reference to natural selection (Morgan, 1909). Morgan's main objections against natural selection can be summarized as follows: selection is only a negative factor and does not provide an account of the rudimentary states of highly specialized structures. If selection of individual variations does not produce anything new and the limits of the species are not transgressed, even the most rigorous selection will not explain evolution. But if there are heritable variations within the bounds of Linnean species, accumulated in a single direction and capable of fixing themselves, one of the most serious criticisms can be surmounted.

Initially, Morgan indeed considers only Devriesian mutations as heritable, but by 1909 he also conceived that individual variations are the outcome of small, heritable differences. He named these wild differences, mutations.

More explicitly, Morgan thought that mutants are often to be considered as extreme modifications or as aberrations, giving the impression that mutational changes involve a great detachment from the original type,[3] and yet small changes are mutations as distinctive as large ones and since differences due to changes in the genes are heritable, evolution takes place through changes in them.

The fact that Morgan kept the term 'mutation' to allude to small Mendelian variations created confusion between the variation through great steps – that according to de Vries would be responsible for the formation of a new species – and small variations taking place within the limits of the species. Morgan includes, in the term 'mutation', any variation of any size and degree that is inherited in a Mendelian mode.

In 1915, Morgan changed his attitude toward natural selection, renouncing the mutational theory of de Vries as an evolutionary explanation (Morgan *et al.*, 1915). The works of Davis (1912) on *Oenothera Lamarckiana* proved that the variations that de Vries took to be mutations were but combinations of hybrid characters of the parental forms that agreed with Mendelian laws. This led many researchers, including

Morgan, to desert the Devriesian position. Muller's discovery of the 'balance of lethals' (1914) also contributed to the definite rejection of de Vries' mutation theory. The concept of 'balance of lethals'[4] was in agreement with Mendelian theory, and it seemed then that Devriesian variations were not as large as de Vries had imagined, that is, they did not take place at the species level. The decline of the mutation theory was an important factor in Morgan's explanation of natural selection in Mendelian terms.

During those years, many discoveries contributed to the acceptance of the Mendelian chromosomal theory of inheritance. Among them were the position effect, genic interaction, recombination and linkage, and, above all, the idea of modifier genes. These modifier genes are a set of genes that can divert the way which another expresses and that can explain the production of small graduations among the Mendelian characters. The strength of the expression of a character will depend on the modifier genes present. Selection will increase or diminish their number producing an infinite number of intermediate forms that explain Darwin's continuous variation.

The chromosomal theory of inheritance, based on studies about the occurrence of spontaneous mutations, established the following facts:

(1) spontaneous mutations occur in one member alone of the pair of genes, but not in both at the same time,
(2) mutation is recurrent, indicating that the process is specific and ordered,
(3) there is a reverse mutation, of basic importance as a genetic criterion to show that mutations do not imply the loss of the original gene.

From this evidence, the idea originated that mutations are changes in genes. According to the gene theory, the 'wild types' are specific genes in chromosomes, relatively stable during long periods of time. For Morgan, there is no evidence that these genes appear as something more than changes in the constitution of old genes or of genes already present. The total number of genes of each species will remain constant. However, this number can be altered by polyploidia or by changes in the chromosome, including the coupling of two chromosomes to form a single one, or the fragmentation of one to become two. For Morgan, these changes are also mutations and are originated in the same way as small individual variations. It is clear, that the subject-matter of evolution will be mutations and the new combinations of already existing characters.

But, what are genes? Once located in chromosomes, the next question was whether they have a real basis, that is, whether they are chemical or physical bodies. The gene had to be studied by observing its effects since it was not possible to do so chemically or physically. The latter analysis would only become feasible when Watson and Crick proposed their double helix model in 1953. For Morgan, however, the problem was about transmission, not about physiology.

Thus the theory of inheritance that Morgan proposed was the combination of two lines: the chromosome theory according to which chromosomes are the transmission agents of inheritance, and Mendelism which emphasized hereditary factors that are segregated and joined in the production of eggs and sperm.

3. MUTATION AND THE THEORY OF THE GENE

One of the most important contributions of Morganian genetics is the proof of the existence of genes. These ultramicroscopic particles influence the entire cell, and play a very important role in determining substances, structures and cellular activities involving the totality of the organism.

The main feature of genes is their ability to self-replicate and self-propagate. Even if genes change they do not lose this ability. Thus, mutation is a general property related to the structure of genes themselves, the same for all of them. This will be of great importance if genes and their variation are to be considered the source and root of organic evolution.

H. J. Muller, Morgan's pupil at Columbia, carried out a conceptual analysis of mutation in 1932. Originally, the term embraced different phenomena having no relation to each other. These had been grouped together because they produced new types appearing suddenly.[5] However, according to Muller, mutation is solely a change in the individual gene. In this sense, mutation will be the alteration of the gene, where alteration means transmissible change, or at least change that can propagate.

The question about evolution, for Muller, was answered thus: the definitions of gene and mutation, indicate that 'gene' is the inheritance unit and 'mutation' the transmissible change. In this sense the mechanism of evolution is thought of as the form, frequency and way of occurrence of mutations.

With this notion of mutation, Muller points to the main attributes of the nature of genes and mutation: the gene is stable and has differential mutability; some genes have a preferential direction in the type of mutation; mutability and its differential direction can also be altered by mutation;[6] the dominance of normal genes is more complete than that expressed by mutants; many mutations have a deleterious effect and mutations with small effects are more frequent than those producing major ones, that is, they are Darwinian in the sense that those small changes that are less deleterious may be selected.

For Muller, these questions served to shape a new theory of mutation that can guide research and inevitably led to the pursuit of the agents producing it. Muller's vision gives birth to a new branch of genetics: mutagenesis.

He began his study using X-rays and discovered that in X-ray-treated germinal cells, real genic mutations occur in high proportions. Different types of mutations are produced: lethal, semi-lethal, dominant lethal, invisible, etc. The mutation rate after radiation is higher in relation to the number of genes involved and, for this reason, mutagenesis became an excellent instrument for the study of gene structure and mutation, both connected with evolution. Muller established that X-ray mutants were similar to mutants produced in nature, and that the effects of X-rays were not unique; they produce point mutations, translocations, inversions and deletions.

It was also shown that an increase in the X-ray dose caused a proportional increase in the production of mutations.

Regarding the characterization of induced mutations Muller concluded that:

(1) The great majority are lethal.
(2) There are more recessive than dominant.
(3) Many of the visible effects are relatively insignificant.
(4) Even if new mutations appear, they do not appear with the mutation frequencies repeated.
(5) All the chromatin regions are disturbed, even if the induced mutations are disturbed in preferential areas.
(6) There is reverse mutation.
(7) The rule is that they are point mutations.

Two conceptual limitations appeared while trying to define point mutations. First, a technique that would show the chemical modifications of the gene was not available. Second, there was an ambiguous classification of point mutations, those that were not small or gross rearrangements. Stemming from this conflict a confrontation was generated between those defending the existence of genetic mutations and those who believed in rearrangements by rupture. With these hypotheses they tried to provide some basis for the understanding of the evolutionary meaning of mutation.

The term 'genetic mutation' is dependent on the particular concept of gene being used. Hence, Stadler in 1954 tried to give a definition of 'gene' based on experimental and theoretical studies of *Drosophila* and corn. For Stadler, clearly, the previous ambiguity in the concept of mutation was derived from the ambiguity in the concept of gene, that is, any definition of mutation presupposes a definition of gene. Stadler proposed an operational definition of gene as the smallest segment of the genetic strand which consistently associates with the specific occurrence of a phenotypical effect. Genetic mutation will then be the change in the gene with visible phenotypical effects and capable of showing reversion. It is obvious that during these years, the concepts of gene and mutation were still unclear.

4. GENE STRUCTURE AND MUTATION TODAY

4.1. Genes and Physiology

Until 1945 the gene was considered the fundamental unit of inheritance, but little was known about its function. Genes could only be identified by mutations that produced phenotypical aberrations. These varied from simple aberrations (eye color) to drastic morphological changes.

During the 30's, Beadle and Ephrussi had tried to associate genes with enzymes. Beadle found that certain reciprocal transplants of some mutants,[7] were genetically related. The hypothesis explaining this phenomenon is that of a chain reaction.[8] From this, Beadle and Tatum (1941) showed that an individual gene was responsible for each of the steps in the chain. This gene specificity corresponds to that of the enzyme, thus displaying the existence of a close relation between them. Later, the hypothesis was named the 'one gene–one enzyme hypothesis'. It claims that each metabolic step is catalyzed by a particular enzyme whose products are accountable to a single gene. A mutation in this gene can nullify the activity of the corresponding protein. Since a mutation is a random event in relation to the structure of the gene, it is highly probable that it may harm the genetic function. In this case, recessive mutations can be explained as the absence of a function because the mutated gene does not produce the required enzyme.

In 1957 Ingram verified that the falciform anemia virus produced a change in the composition of hemoglobin amino acid and stated that genes are burdened with the task of controlling protein structure. With these studies, the original hypothesis of 'one gene–one enzyme' was replaced by a more precise one, 'one gene–one polypeptide'. Thus, a gene would be the consequence of nucleotides which codes for a polypeptide chain or for a protein dominion, and a mutation is the phenomenon altering the expression of the gene to yield the alteration of the original protein function.

Thus, if by 1901 Hugo de Vries conceived a gene as a character and, later, in 1914, Morgan thought of it as a factor, with Beadle and Tatum, in 1941, it became possible to define the gene as the entity capable of producing a protein with a certain function. Gene problematic had progressed because of theoretical and methodological advances in biology.

4.2. Gene Structure and Evolution

The discovery of multiple alleles[9] by Morgan in 1913 posed the problem of the independent origin of the alleles, and of whether these factors formed an integrated series indeed formed integrated segments of the chromosomes where no recombination took place. Hence, the origin of multiple alleles is the change in each place of this lineal fragment thus producing the different alleles in an independent way.

This problem was temporarily abandoned because experiments were not planned to detect the allele recombination in the same series. Only a great number of discoveries of allelic series caused the problem to be reconsidered.

Oliver (1940) reported a reversion to the wild type associated with recombination in *Drosophila melanogaster*. This reversion might be due to unequal recombination or it might be explained under Bridge's hypothesis (1917) according to which different loci are involved in the expression of two mutants. This hypothesis of Bridge's repetition proposes the evolution of genes by tandem duplications with a subsequent independent evolution.

Lewis (1941) reports another case with the dominant mutant Star and its recessive allele. He found two complementary types of recombination and the phenotypes for the two arrangements. He named this alleles position pseudoalleles because a difference existed with respect to the cis and trans arrangements. That is, the alleles can be distinguished by their expressions and by the cis–trans difference. This was interpreted as though they had appeared by duplication and were in a changing process: their functional resemblance would be maintained as a positional effect.

The presence of cis–trans difference indicated the existence of related genes (position pseudoalleles). In the trans case each chromosome would block the biochemical path in one or two of the possible points, preventing the formation of a certain substance. In the cis arrangement, a chromosome with two normal alleles would complement the biochemical step, and therefore, that arrangement would have the tendency to present a more normal phenotype than the trans one. The idea of two divergent genes, once identical to each other in their functional result, is very attractive and persuasive since it offers a process required for a general theory of gene evolution. Furthermore, the development of sequence steps at this level, is only possible in the case of genes that were once identical (Lewis, 1951).

Pontecorvo (1955) made a distinction between the Lewis effect (position pseudoallelism) and complementation. For Pontecorvo, the gene should be defined in three senses: as a recombination unit, as a mutation unit and as a physiological unit. The Lewis effect should be explained as a cis–trans phenomenon that seemed to be a particular case of complementation, in the third sense. Pontecorvo thinks that the distinction between the Lewis effect and complementation is a consequence of spatial organization, i.e. of distance. Mutable places near each other originate alleles with a Lewis effect, while far places yield complementation: thus there is a relation that links the distance between alleles to the complementation degree.

4.3. Fine Structure

In bacterial genetics the transduction test allows for separation of auxotrophic mutants[10] in well-defined groups. This means that the members of each group are allelic and the presence of transduction inside a group is explained in the same manner as the low frequency of recombination between pseudoalleles. This has suggested that the locus extends over a chromosome section and that the changes taking place in different regions of this section give place to different alleles. It also indicates that the regions within this section are separable and recombinable with homologous regions within the locus of another chromosome.

Benzer in 1955, studying the rII region in the bacteriophage T4, poses the question of how closely this resolution level reflects the molecular limits of the genetic material. The mapping procedure was based on the frequency of recombination of the different crossings. He characterized in molecular terms the extension and size of the units of recombination, mutation and function. In these first experiments the magnitude of the recombinational unit was calculated in a dozen nucleotide pairs and the mutational unit in as many as hundreds of nucleotide pairs.

A more detailed analysis of the rII region organization of T4, revealed that this region is divided into two further regions (A and B) where the classical genes are organized in a dimensional arrangement into divisible chromosomes by recombination (Benzer, 1956). This allowed him to define the gene in a new way: recon is the smallest recombination unit in a one dimensional arrangement, that is exchangeable but not divisible through genetic recombination; muton is the mutational unit, understood as the smallest element which, altered, gives place to a mutated form of the organism; and cistron is the functional unit that can be genetically defined[11] through the Lewis cis–trans test.

The notion of the gene as the functional, mutational and recombinational unit was shown to be obsolete by those studies of fine structure. Benzer (1957) using new conceptual terms simplified the gene polemic. The classical gene was divided into cistron, recon and muton; and these did not coincide because their measurements were different. The 'gene' that produces a specific protein or a message was called 'cistron'; the smallest segment capable of recombination was called 'recon', and the gene, defined as the smallest mutable segment, was named 'muton'. Of these terms, 'cistron' is the most widely used, especially in microbic systems as a synonym for gene, but with a clear connotation that makes reference to the fine structure found by Benzer in the rII region of T4.

4.4. *Molecular Explanation of Mutation*

Muller (1930) recognized and verified the highly localized quality of point mutation, and proposed that it results from a chemical or physical change of the gene.

The Watson and Crick model (1953) and Benzer's studies of the fine structure of the rII region of the T4 bacteriophage provided the molecular base for the mutagenesis theories that attempted to solve Stadler's paradox. This stated that a mutation is, first, a change of unknown nature and second, that it is impossible to discriminate between point mutations and rearrangements.

Watson and Crick's model suggested that a pair of nucleotides, if subtracted (deletion) or replaced, would cause a point mutation. Benzer demonstrated that there is a relation between the DNA molecule and the mutable places in the fine structure of the genetic map. Freese (1959) mapped the mutations produced in the T4 phage by an analogue base, 5-bromouracil, and observed that the agents used required a DNA duplication in order to produce a mutation. Then he proposed two different types of mutation: transitions and transversions.

Transitions are errors in matching that cause the replacement of a purine by another purine or a pirimidine by a similar one, causing a *missense* mutation originating a protein with a replaced amino acid.

Transversions are the replacements of a purine by a pirimidine, and *vice versa*, caused by acridine compounds. However, Brenner (1958) had demonstrated that acridines produced minuscule intragenic deletions or duplications no larger than a pair of nucleotides. Brenner proposed that the acridine compounds are interpolated in the DNA molecule causing an inexact copy of the sequence by the insertion of one or more bases. These types of mutations were called 'frameshift mutations' and the mutants have the peculiarity of not producing a protein similar to that of the original or wild type.

There are difficulties in the interpretation of the mutagenic activity of other compounds, such as nitric acid, formaldehyde or alchilant agents, whose reactions with DNA are complex. Nevertheless, Freese and Brenner's studies about the molecular explanation of point mutations have been stimulating for research on higher organisms.

4.5. Split Genes

One of the most outstanding discoveries of molecular biology is that genes are split or fragmented in a series of alternating regions encoding discrete domains of proteins (exons) and interpolated ones (introns) that do not encode but that are transcribed with the former to form a single RNA molecule. Through a process called 'RNA splicing' the introns are removed from the RNA generating a functional messenger RNA (RNAm) that contains the required information for protein synthesis. Thus, exons are the evolutionary and functional units encoding proteins. The idea is suggested that many of the present genes could have originated from duplication. However, not all the genes can have new functions: a duplicate gene can accumulate deleterious mutations and become non-functional (pseudogene). A pseudogene is a DNA segment that shows homology with a functional gene, but that contains flaws like senseless mutations, or frameshift ones, that bar the elaboration of the functional product (a polypeptide). Present evidence indicates that pseudogenes have properties that lead us to think that they are derived from the dysfunction of the duplicate genes (Li, 1983). In this way, pseudogenes appear as mutants that generate termination codons between exons, changes in the processing of RNA, or interruption sequences, producing changes in the DNA frame reading. As selection acts upon functional exons and not upon pseudogenes or introns, the unit of selection is the exon. This is the current concept of the gene at a molecular level.

4.6. Transposition

Another important discovery of the 50's is the existence of small segments of genetic material that can move within the genome. This was discovered by Barbara McClintock in the 1940s while working with corn. She found out that the coloring irregularities in the seed were caused by mutations at either an early or at a late stage of development. The number of patterns of coloring tiles reflect the rate at which these mutations take place. Each seed has a characteristic mutation rate that does not vary during the life cycle. McClintock came to this by assuming that something was controlling the mutation rate. This regular rate showed the regularity of development to be linked to genetic events and not, as was believed, to embryological ones, where mutation would have two direct effects: first, it would allow the tracing of the history of cellular differentiation, and second, it allowed establishment that these patterns were not produced at random.

The transposition or movement of small DNA segments is a two step process: freeing a chromosome element from its original position (dissociation) and inserting it into a new position. McClintock called the first family of movable genetic elements (or controlling elements) the Dissociator-Activator Family (Ds-Ac).

McClintock also proved that mutations produced by this family impact the level of genic expression, the structure of the genetic product and the development of the expression as a consequence of changes in the sequence inside and outside the transcriptional unit at a locus. These studies indicate that mutations by transposable elements, like Ds-Ac, modify the structure and expression of the locus in different ways, for example, they can reduce the transcriptional level, alter the development timing, and change the primary structure of the transcriptional unit. This ability to bring about mutations is manifest in corn in the uniform size of the somatic sectors coming from the mutated cell.

The controlling elements, later on transposons, insertion sequences, extrachromosomal DNA, virus and bacteriophages, suggest the existence of genetic determinators that change their position in the genome, altering the gene expression in which they have been inserted. In recent years many studies about transposition have been made (Shapiro, 1983) and similarities with other organisms have emerged.

The idea that the genome is not a static entity, but rather a complex structure in dynamic equilibrium where all the movable elements have the same structural organization, is already accepted. Our conception of genes seen through this movable element is different and radical; Keller proposed in 1983 that mutation is a process neither at random nor directed but influenced by the organism itself.

5. CONCLUSIONS

Ever since the introduction of the term 'mutation' by Hugo de Vries in 1901 it has become increasingly precise. The discovery of *Drosophila* mutants by Morgan[12] was the key to understanding that genetic mutations are the source of organic evolution. The question of whether mutations can be deliberately induced and quantified, was answered affirmatively. The intense activity in mutation research brought as a consequence the creation of a new field in Genetics: mutagenesis. This became a theory and a refined explanation of genetic evolution, corroborating the view that mutation is responsible for the creation of new variants in evolution.

The search for genes as a physical corpuscle was justified by the results obtained by Muller using X-rays. His conception of the gene as a material particle had an immediate effect: if genes are 'real' (material), we can hit them with electrons and measure their size. This concept of gene, born from *Drosophila* research, is radically different from Darwin's gemules, de Vries' pangenes, etc. It was born from the operative thought launched by Morgan in 1914 with his theory of the gene later confirmed by cytogenetic methods. The experimental projects designed by Muller were determinant of this new conception of genetics and of experimentation as such. Briefly, the study of mutation as an instrument for studying genes was beyond the Devriesian idea of mutation conceived as gross phenotypical alterations: there is a strong relation between genes and mutation, and this relation is an evolutionary one.

The aim of this paper has been to establish the connections between a scientific discipline, genetics, and the history of science. That is, following Canguilhem (1975), to establish the genesis and the lineage of the concepts present in genetics that, when conceived as answers to problems, have led to a sequence of theories in which we can recognize "the persistence of a problem inside the theory that believes it has found its solution" (Lecourt, 1972). The history of genetics, viewed as a conceptual history, reveals a succession of theoretical and practical junctions that constitute scientific knowledge. Experimentation reveals the technical possibility of dealing with old problems (such as variation), providing them with novel solutions. Thus, once more, we can reaffirm the idea that scientific theories and concepts have a history and that they are generated within a culture and a theoretical horizon that allow them to come into being.

From the inception of genetics, the concept of the gene has been clarified. Gene localization in chromosomes and its way of transmission were the immediate problems attacked by the forefathers of genetics. Its modification by mutation constituted a second phase, giving way to the development of new techniques and to the use of new organisms in order to explain the structure, size and function problems. The molecular phase, and with it codification and regulation models, have made the gene a precise concept.

NOTES

1 For example, the effects of mutagens like X-rays on treated material, are well known.

2 Individual or continuous, and singular, discontinuous, or 'sports'.

3 As in Darwinian leaps, deemed to be large alterations that deharmonize individuals

from the environment to which they are adapted.

[4] Lethal genes that remain within a population because their lethal nature only appears in homologous condition.

[5] Recombination, abnormal distribution of chromosomes and point mutations.

[6] That is, the frequency of mutation is genetically controlled, and can be altered by mutation elsewhere.

[7] In the eye color in *Drosophila*, vermillion, claret and cinnabar.

[8] Where a mutated gene produces the chain interruption (substance CA + substance V + substance CN), which modifies the normal coloring of the eyes, yielding thus any alteration in the process visible.

[9] I.e. alternative forms of a gene.

[10] Auxotrophic mutants are those that lose the capacity of growth in a determined environment, but that are viable in a supplemented environment.

[11] Unconnected with the biochemical information.

[12] Many of which are alleles of other mutants.

REFERENCES

Ayala, F. J. and J. K. Kiger Jr. 1980. *Modern Genetics*. Benjamin/Cummings Pub. Co. Inc.

Beadle, G. W. and E. L. Tatum. 1941. 'Genetic control of biochemical reactions in Neurospora'. *Proc. Natl. Acad. Sci. USA*, **27**: 499–506.

Brenner, S., S. Benzer and L. Barnett. 1958. 'Distribution of proflavin induced mutations in the genetic of fine structure'. *Nature*, **182**: 983–985.

Benzer, S. 1955. 'Fine structure of a genetic region in bacteriophage'. *Proc. Natl. Acad. Sci. USA*, **41**: 862–863.

Benzer, S. 1956. 'Genetic fine structure and its relation to the DNA molecule'. *Brookhaven Symposia in Biology*, **8**: 3–5.

Benzer, S. 1961. 'Fine structure mapping of the rII region of bacteriophage T4'. *Proc. Natl. Acad. Sci. USA*, **47**: 403–415.

Benzer, S. 1961. 'Genetic fine structure'. *Harvey Lectures*, **54**. Ac. Press Inc., N.Y.

Canguilhem, G. 1975. *Etudes d'histoire et de philosophie des sciences*. J. Vrin, Paris.

Davis, B. M. 1912. 'Genetic studies in *Oenothera* III'. *Am. Nat.* **56**(547): 377–427.

De Vries, H. 1901–1903. *Die Mutationstheorie*. Veit & Co, Leipzig.

Dobzhansky, T. 1970. *Genetics of the Evolutionary Process*. Columbia University Press.

East, E. M. 1929. 'The concept of the gene'. *Proc. of Inter. Cong. of Plant Sci.*, **1**: 889–895.

Freese, E. 1959. 'On the molecular explanation of spontaneous and induced mutations'. *Brookhaven Symposia in Biology*, **12**: 63–75.

Herskowitz, I. H. 1962. *Principles of Genetics*, 2nd ed. McMillan Pub. Co. Inc.

Johannsen, W. 1909. *Elemente der exakten Erblichkeitslehre*. G. Fischer, Jena.

Keller, E. F. 1983. *A Feeling for the Organism*. Freeman & Co., N.Y.

Lecourt, D. 1972. *Pour une critique de l'épistémologie*. Maspero, Paris.

Lewis, E. B. 1941. 'Another case of unequal crossing over in *Drosophila melanogaster*'. *Proc. Natl. Acad. Sci. USA*, **27**: 31–34.

Lewis, E. B. 1951. 'Pseudoallelism and gene evolution'. *Cold Spring Harbor Symposia on Quantitative Biology*, **16**: 159–174.
Li, W. H. 1983. 'Evolution of duplicate genes and pseudogenes'. *Evolution of Genes and Proteins*. Nei, M. & R. K. Kohn (eds.). Sinover Ass. Inc., Sunderland, Mass.
McClintock, B. 1947. 'Cytogenetic studies of maize and *Neurospora*', *Carnegie Inst. Wash. Year Book*, **46**: 146–152.
Morgan, T. H. 1909. 'For Darwin'. *Popular Science Monthly*, **70**: 367–380.
Morgan, T. H. 1910. 'Chromosomes and heredity'. *American Naturalist*, **44**: 449–496.
Morgan, T. H. 1917. 'The theory of the gene'. *American Naturalist*, **51**: 513–544.
Morgan, T. H. 1922. 'On the mechanism of heredity'. *Proceedings of the Royal Society of Biology*, **94**: 162–197.
Morgan, T. H., A. H. Sturtevant, H. J. Muller and C. B. Bridges. 1915. *The Mechanism of Mendelian Heredity*. Henry Holt and Co., New York.
Muller, H. J. 1914. 'The bearing of the selection experiments of Castle and Phillips on the variability of genes'. *American Naturalist*, **48**: 567–576.
Muller, H. J. 1917. 'An Oenothera-like case in *Drosophila*'. *Proc. Natl. Acad. Sci. USA*, **3**: 619–626.
Muller, H. J. 1927. 'The problem of gene modification'. *Proc. of the 5th International Congress of Genetics*. Berlin 1927. *Zeitschrift für induktive Abstammungs- und Vererbungslehre Suppl.*, **1**: 234–260.
Muller, H. J. 1932. 'Further studies on the nature and causes of gene mutations'. *Proc. of the 6th Inter. Congress of Genetics*, **1**: 213–255.
Muller, H. J. 1938. 'The present status of the mutation theory'. *Current Science*. March.
Oliver, C. P. 1940. 'A reversion to wild type associated with crossing-over in D. melanogaster'. *Proc. Natl. Acad. Sci. USA*, **26**: 452–454.
Pontecorvo, G. 1955. 'Gene structure and action in relation to heterosis'. *Proc. of the Royal Society of Biology*, **144**: 171–177.
Shapiro, J. 1983. *Mobile Genetic Elements*. Academic, New York/London. 688 pp.
Stadler, L. J. 1954. 'The gene'. *Science*, **120**: 811–819.
Watson, J. D. 1976. *Molecular Biology of the Gene*, 3rd ed. Menlo Park, CA, W. A. Benjamin Inc.
Watson, J. D. and F. H. C. Crick. 1953. 'Molecular structure of nuclei acids'. *Nature*, **171**: 737–738.
Watson, J. D. and F. H. C. Crick. 1953. 'Genetical implications of the structure of deoxyribonucleic acid'. *Nature*, **171**: 964.

ELIA NATHAN BRAVO

GALILEO'S REVOLUTION: THE USE OF IDEALIZATIONAL LAWS IN PHYSICS*

It is generally accepted that Galileo was the first physicist to introduce the systematic use of mathematics into mechanics, although a mathematized statics and optics were already known by the Middle Ages. Furthermore, some historians of science, notably Koyré (for example in 'Galileo and Plato', [9], pp. 150–179), consider this mathematization of physics to have been one of Galileo's greatest revolutionary contributions. Since this question has been relatively well discussed, what I would like to investigate in this paper is a related issue, the use of idealizational laws by Galileo. I will try to show that Galileo was able to introduce mathematical laws into physics by making them idealizational. Furthermore, I think that by studying Galileo's use of idealization, we can understand better his revolutionary contributions – namely, what is it that differentiates in a radical way the kind of theory that he developed from the kind of theory that Aristotelian physics was and why experimentation came to be a fundamental tool for the development of classical physics.

This paper is divided into two sections. In the first one, we shall study what is an idealizational law, using as an example of it Galileo's law of free fall. In the second section, we shall try to show briefly how physics was deeply affected by the introduction of idealization, so that it came to be a different kind of theory from the one it was before Galileo.

I

Galileo was the first to introduce idealizational laws into physics (although in statics this had already been done by Archimedes). Since then, the use of idealizational laws, concepts or theories has become frequent, both in physics (examples of this can be found in [1], [10] and [15]), and in the social sciences ([1], p. 158; [13]). We shall study here only idealizational laws. The thesis we will argue for in this paper is that Galileo introduced a special type of law into physics, which we call

S. Ramirez and R. S. Cohen (eds.), Mexican Studies in the History and Philosophy of Science, 109–128.

'idealizational'; this does not mean that all laws of physics, Galilean or any other, are idealizational – indeed, some are non-idealizational or factual – nor does it mean that all theoretical laws are idealizational.[1]

To determine the characteristics of idealizational laws and the way they function we shall study Galileo's law of free fall. This law, Galileo stated, is constituted by two claims:

(1) The fall of a body in the vacuum (i.e., where the resistance of the medium is zero) is uniformly accelerated, that is, the increment of speed is proportional to the increment of time ([8], pp. 161, 163, 170; [7], p. 221). From the thesis that speed is proportional to time, Galileo deduces that the distance is proportional to the square of the time interval employed in traversing this distance, that is to say, the distances traversed in each unit of time bear to one another the same ratio as the series of odd numbers beginning with number one ([8], pp. 174–176; [7], p. 222); and also he mathematically deduces other properties.
(2) All bodies, independently of their weight, dimensions and nature, fall in the vacuum with the same acceleration ([8], p. 72), that is, the acceleration of free fall is a universal constant.

Let us now consider idealizational laws. We shall suppose that a law includes not only a claim about certain entities or relations among entities (or their properties) but also the conditions under which such claim is satisfied. Let us take as an example the Galilean law of free fall:

> Given any bodies 'x' and 'y', if 'x' and 'y' move in the vacuum (i.e., in a medium in which the friction that acts upon 'x' and 'y' is equal to zero and the upward-thrusting force of the medium that acts upon 'x' and 'y' is equal to zero), then the median speed with which 'x' falls is equal to the median speed with which 'y' falls.

This law is usually considered to be *idealizational* because it contains at least one *known ideal condition*, that is, a condition which is not satisfied in real situations (although some real situations can be approximated to the ideal conditions). The ideal condition in this law is that the bodies should move in the vacuum ([10], pp. 15–17; [1], p. 258]. In Galileo's case the ideal character of this vacuum condition is somewhat problematic, since Galileo argued correctly that motion in the vacuum is possible and that, furthermore, a vacuum can be produced with a certain instrument ([8], p. 67). This problem can be solved in two different ways.

(a) Galileo was conscious that in real situations outside the laboratory, the vacuum condition is not satisfied, and because of this, with respect to the 'normal' sensible world, which is the one science seeks to explain, it can be considered as an ideal condition. Thus, for example, to Simplicio's (an Aristotelian) objection that because the resistance of the medium cannot be prevented, the law of free fall (which is claimed for bodies that fall in the vacuum) is not true for the real world, Galileo does not answer by claiming that there can be a vacuum in the sensible world, but by expounding certain aspects of the idealizational method ([8], p. 251, cf. also p. 74).

(b) Another solution is to consider that the fundamental characteristic of idealizational laws is not that they contain ideal conditions (although in many idealizational laws it is evident that the conditions are ideal) but that some of their conditions establish that there are certain variables whose value is zero, even though it *is known* that in the normal real situations the *value of such variables is not zero*. Thus, the Galilean law of free fall is idealizational because it considers the value of the medium's resistance to be zero; it is as if it did not exist, even though Galileo knows, and he says it many times, that in the normal, natural, situations (outside the laboratory) the value of the resistance is not zero; in other words, the law that describes the fall of bodies is idealizational because it does not take into account the resistance that the medium offers to a falling body, although it is known that the resistance does influence the normal fall of the bodies. (Even though I prefer this second characterization of an idealizational law, I shall use any of them interchangeably).

Not every law is idealizational, there are also factual laws, where conditions can be satisfied by normal situations. An example of a factual law is the following:

> If a body falls in a resisting medium, then $V = c(\frac{G-G_m}{G})$ (where 'V' is the medium speed, 'c' is a constant of proportionality, probably the constant of acceleration in the vacuum, 'G' is the gravity or specific weight of the body that falls and 'G_m' the specific gravity of the medium).[2]

Thus, the basic difference between factual laws and idealizational ones is the satisfiability, or lack of it, of its conditions by real objects;

or in other words, that the variables that influence the phenomenon described by the law either do or do not have a value of zero.

Given the characteristics of idealizational laws we may wonder how do laws, which are only true for ideal cases, relate to, or are applicable to, the real world. Clearly this relation has to be indirect and complex. It is achieved through two procedures: factualization and experimentation.

Before examining Galileo's conception of these, it is necessary first to examine Galileo's view of the problem of idealization.

The words that Galileo used are 'abstract' and 'concrete'. With the term 'abstract' Galileo refers both to the mathematical or quantitative aspect of a law, and also to the fact that a law contains ideal conditions. Thus, for example, he considers the problem of how mathematics, which is abstract, can tell us something about what is concrete or real ([7], pp. 203–208). On the other hand, he claims that the conclusions which are proven in an abstract way about uniformly accelerated motion – for example, that $d \propto t^2$, given that $v \propto t$ – will be different when they are applied to concrete cases – for example, to the real fall of an object – given the resistance of the medium ([8], p. 251). This claim refers to the fact that the law of free fall is abstract in the sense of being idealizational, that is, of containing the ideal condition of the vacuum. With the term 'concrete' Galileo refers to what is real, i.e. to the sensible reality which is apprehended or described in accordance with a certain theoretical or philosophical position, which in Galileo's case is mechanism.

I believe that although it can be correctly said that mathematized laws and idealizational laws are abstract, they are abstract in different senses. It is usually claimed that a concept, law or statement is abstract if it does not take into account certain aspects of reality.[3] (This note should be read carefully because it is quite important.) Thus, the reason Galileo might have had to consider mathematized physical laws as abstract is that they only take into account functional relations among qualities, leaving aside all considerations about any other characteristics of the object or objects that have the qualities in question: for example, the law of free fall does not take into account the kind of thing that the falling object is, the place from where it falls or to where it falls, etc. On the other hand, the law of free fall can also be abstract in the sense that it does not take into account, in the functional correlation, a factor which is known to affect such correlations, namely, the resistance of the medium. Therefore, although idealizational and mathematized laws are abstract, the character of the abstraction they contain is different. In the case of mathematized laws, it is presupposed that the variables that are left aside do not affect the phenomenon under consideration, while in the case of idealizational laws *it is known* that the variables which are left aside *do influence* the phenomenon being described by such laws.

We have tried to distinguish between mathematization and idealization. Let us now see how they are related. According to Nowak, Galileo was able to introduce mathematics into physics because his physics contains idealizational laws ([13], p. 7). Nowak explains why idealization is a requirement for mathematization in the following way:

> The application of mathematics directly to the description of phenomena as they appear to us in our experience is impossible – too many influences act there. Some of these influences – with the help of idealizational presuppositions (e.g. the elimination of the resistance of the medium) – have to be eliminated, and only then can simple quantitative relations be established among physical entities ([13], p. 7).

I consider Nowak's claim that the mathematization of physics necessarily requires idealizational presuppositions, that is, that mathematization depends upon idealization, to be incorrect. Let it suffice here to look at the example of the so-called peripatetic law of motion. This law states that $V = \frac{F}{R}$ (where 'V' is the speed, 'F' the force, and 'R' the resistance). This law was never considered idealizational by those medieval thinkers who held it, nor by Aristotle,[4], since they did not establish that this law contains ideal conditions; neither did the critics of this law consider that it was true for some ideal case.

In the same way as mathematization does not in principle require idealizational presuppositions, clearly idealization does not require mathematization – for example, Marxist theory contains idealizational concepts (the concept of the capitalist mode of production is one of them) but is not mathematized. Therefore, we can say that the scientific procedures of mathematization and idealization are different, and furthermore, are not necessarily interrelated.

Nevertheless, in Galileo's case both procedures are related in an intimate way. The mathematical treatment of free fall and of projectile motion was carried out by Galileo under the idealizational presupposition that these motions take place in a vacuum. As Nowak points out, the mathematization of a law may require elimination of some of the many influences which are present in the actual phenomenon, and one way to achieve this elimination is through the use of ideal conditions.

To sum up, the procedures of idealization and mathematization are, in principle, different. In Galileo they are intimately related since it is due to the use of idealizational presuppositions that he was able to mathematize laws; this led Galileo to confuse or identify these two procedures, and to consider them as both being cases of abstraction (in the same sense). In spite of this confusion, Galileo's answer to the problem of the applicability of abstract claims to concrete cases is interesting and correct, and therefore we shall examine it.

Galileo's view is that there are two ways in which abstract statements are related to the concrete cases. The first is through a process of factualization (cf. [10], pp. 23–25), or theoretical manipulation which enables one to apply abstract statements to concrete cases or normal real situations. The second is through experimentation, which enables one to apply abstract statements directly (that is, without a process of theoretical manipulation) to concrete cases (which in this instance are artificial, or laboratory, real situations), and which would make possible some kind of verification of the abstract statement. We shall examine each one of these procedures in more detail.

In the *Two World Systems*, Simplicio (an Aristotelian) argues against the introduction of mathematics into physics by claiming that, given the imperfection of matter, the conclusions that are reached in the abstract cannot be applied to the concrete. Thus, for example, a line which is tangent to the surface of the Earth in fact does not touch it at only one point as claimed by geometry. To this objection, Salviati (Galileo's spokesman) answers that it is not the case that the concrete is different from the abstract since, to take the last example again, a tangent does not touch the surface of the Earth at only one point because the Earth is not a perfect sphere. And, he continues:

> But I tell you that even in the abstract, an immaterial sphere which is not a perfect sphere can touch an immaterial plane which is not perfectly flat in not one point, but over a part of its surface, so that what happens in the concrete up to this point happens the same way in the abstract. (...) Just as the computer who wants his calculations to deal with sugar, silk and wool must discount the boxes, bales and other packings, so the mathematical scientist (*filosofo geometra*), when he

> wants to recognize in the concrete the effect which he has proved in the abstract, must deduce the material hindrances, and if he is able to do so, I assure you that things are in no less agreement than arithmetical computations ([7], p. 207; cf. also [8], p. 170).

Galileo's view is that there is no radical difference between mathematics, i.e. the abstract, and the real world, i.e. the concrete;[5] the concrete fits the abstract if one adds to the abstract those factors that mathematics had abstracted, or *vice versa*, if one subtracts from the concrete those factors that mathematics had abstracted. This thesis, as Galileo states it, is quite interesting, but too general to help us in particular cases.

In his *Two New Sciences*, he specifies how abstract statements (abstract because they are idealizations), relate to concrete situations: through the reintroduction of the abstracted factors. In Galileo's words:

> I grant that these conclusions proved in the abstract will be different when applied in the concrete and will be fallacious to this extent, that neither will the horizontal motion be uniform nor the natural acceleration be in the ratio assumed, nor the path of the projectile a parabola, etc. (...) When we wish to apply our proven conclusions to distances which, though finite, are very large, it is necessary for us to *infer, on the basis of demonstrated truth, what correction* is to be made for the fact that our distance from the center of the earth is not really infinite [ideal condition of the abstract statement] (...) and having discovered and demonstrated the theorems [of free fall, projectile motion, etc.], in the case of no resistance, to use them and apply them with such limitations as experience will teach ([8], pp. 251–253; emphasis mine).

Thus, Galileo claims that in order to apply idealizational laws to real situations they must be corrected, that is, the ideal condition have to be replaced by real or factual conditions (process of factualization). This correction is made on the basis of theoretical principles, or on the basis of principles learned by experience.

The law of free fall (all bodies fall in the vacuum with the same acceleration) is corrected by introducing the resistance of the medium. Although Galileo is not too precise about the way by which this correction is made, he makes some remarks which are quite specific.

He claims that *experience* teaches us that the resistance of the air acts in two ways: on the one hand, it offers more resistance to the bodies which are less dense than to the bodies which are more dense, and on the other hand, it offers more resistance to a body which moves more quickly than to that same body if it moves more slowly ([8], p. 253). Although this empirical principle does help us to understand in a qualitative way why bodies in air do not fall with the same acceleration, it does not help us to determine their speed quantitatively or numerically.

In another section of the *Two New Sciences*, Galileo appeals to a hydrostatic *theoretical* principle to calculate the speeds of bodies that fall in a resisting medium, that is, to make the necessary corrections. This principle is the following: "the medium diminishes the weight of any substance immersed in it by an amount equal to the weight of the displaced medium; so that air in air loses all its weight" ([8], p. 81). On the basis of this principle, Galileo claims that the real speed of a body that falls in a resisting medium is calculated "by observing how much the weight of the medium detracts from the weight of the moving body" ([8], p. 75).[6] In other words, the formula to calculate the medium speed of a body that falls in a resisting medium would be the one that we mentioned before: $c(\frac{G-G_m}{G})$.

Now let us look at the second way in which abstract statements, or more precisely, idealizational statements, are related to the concrete, real, world, namely, experimentation. Take as an example the law that bodies fall with uniformly accelerated motion in the vacuum. From the principle that in such motion, speed is proportional to the time, Galileo derives the theorem that in uniformly accelerated motion the distance traversed is proportional to the square of the time. Then he confronts this theorem with observed experience, and since he finds that they are in agreement, he concludes that the principle $V \propto t$ is true, and that therefore, free fall is a case of uniformly accelerated motion (a method which today is called 'hypothetico-deductive method') ([8], p. 160).

The way in which the theorem $d \propto t^2$ confronts observational experience is through experimentation.

The experiment that Galileo made (or at any rate suggested) consists in working with inclined planes along which bodies roll. He chose to work with balls that roll on inclined planes instead of objects that fall because in the first case the fall is slower, and since he did not have very sensitive clocks, he had thus a better opportunity to determine more precisely the time of fall ([8], p. 84; [7], p. 28). The experimental situation is the following:

> A piece of wooden moulding or scantling, about 12 cubits long, half a cubit wide, and three finger-breadths thick, was taken; on its edge was cut a channel a little more than one finger in breadth; having made this grove very straight, smooth, and polished, and having lined it with parchment, also as smooth and polished as possible, we rolled along it a hard, smooth, and very round bronze ball. Having placed this board in a sloping position, by lifting one end some one or two cubits above the other, we rolled the ball, as I was just saying, along the channel, noting, in a manner presently to be described, the time required to make the descent. We repeated this experiment more than once in order to measure the time with an accuracy such that the deviation between two observations never exceeded one-tenth of a pulse beat. (...) in such experiments, repeated a full hundred times, we always found that the spaces traversed were to each other as the squares of the times, and this was true for all inclinations of the plane ([8], p. 178–179).

Let us emphasize the central points of this experiment

(1) Galileo emphasizes that the channel is very straight, smooth and polished, and that the ball should also be very round, polished and smooth. In the *Two World Systems* he also uses the (imaginary) experiment of the inclined plane to 'prove' the principle of inertia, and he says:

> Remember that I said a perfectly round ball and a highly polished surface, *in order to remove all external and accidental impediments*. Similarly I want you to take away any impediment or the air caused by its resistance to separation ([7], p. 146, my emphasis; cf. also [8], p. 253).

Let us remember that for Galileo the law of free fall ($V \propto t$) is idealizational because it can only be satisfied under the ideal condition of the vacuum ([8], p. 170), as is also the case in the principle of inertia. Therefore, the law, or its consequences ($d \propto t^2$) cannot be directly confronted with sensible experience, but rather this experience should occur within (real) conditions that are as similar as possible to the ideal conditions. As the ideal condition in this case is that there should not be any resistance by the medium, the real condition under which the fall of the balls along inclined planes occurs, must have characteristics which reduce air resistance to the maximum possible. This, I believe, is a fundamental point. Experimentation is a situation in which the conditions under which a phenomenon happens are controlled. One reason why the control of conditions may be required is the necessity to adjust the phenomena to the theory, and more precisely, the necessity of approximating the real conditions under which a phenomenon occurs to the ideal conditions stated by the law. From these considerations we may conclude that a fundamental reason why we must have recourse to experimentation, to the creation of artificial situations, and not to mere sensed observations in experience, is the necessity to relate abstract (i.e., idealizational) laws to concrete phenomena.[7]

Some historians of science have pointed out that experimentation has the function of connecting abstract or idealizational laws with concrete phenomena. For example, I. B. Cohen claims that

> it was Galileo who first put the scholastic laws of motion [in particular, the law or definition of uniformly accelerated motion] to the test of experiment and proved that they could be applied to the real world of experience. (...) no one else in that 300-year interval had had the insight to see how to relate such abstractions to the world of nature ([3], p. 112).

(2) Galileo emphasized that he repeated the experiment many times in order to obtain precise or exact results. That there is the requirement that results should be precise is due to the fact that the law which is to be related to the concrete world is abstract in the sense of being mathematized, so that what is to be related are numbers obtained theoretically with numbers obtained experimentally.

(3) A noteworthy characteristic of the kind of experimentation that is carried out in relation with mathematized laws, is the use of instruments that make possible the creation of artificial situations, or that make possible measurements which are each time more precise.

To sum up, the way in which abstract laws are related to, or are applied to, the concrete world is through two procedures: (1) A procedure of theoretical manipulation by which ideal conditions are replaced by real conditions; and/or (2) a procedure of experimental manipulation through which real conditions are approximated to the ideal conditions.

We can now ask why, if the application of idealizational laws usually requires factualization, does the theory contain idealizational laws, and not such laws in a factualized form (which presents the advantage that they are directly applicable to reality)? The answer to this question is somewhat complex. Among other things, it will show us what sort of theoretical function idealizational laws may have.

First, in the example of the factualization of the law of free fall it can be clearly seen that the replacement of the ideal variables by real ones (or in other words, the replacement of variables whose value is zero by variables whose values are real numbers), alters the functional relation that is stated as the consequence of the law. Now, it is not always possible, or convenient, to substitute real conditions for the ideal conditions, for two reasons.

(a) It is possible that a functional relation which includes real variables cannot be found. This happened to Galileo:

> As to the perturbation arising from the resistance of the medium this is more considerable and does not, on account of its manifold forms, submit to fixed laws and exact descriptions ([8], p. 252).

(b) It is possible that the factualized law turns out to be very complex, and for this reason it may be convenient to keep a simple idealizational law in the theory. Thus, it is partially for pragmatic reasons that a theory has idealizational laws.

Second, the use of idealizational laws, especially when they cannot be replaced by factualized laws, has a theoretico-heuristic function, since they make possible the organization, even though in a crude way, of the empirical data that are available. This organization of the data may help further refinement of the theory, or of the instruments or of the manipulative techniques that are part of the experimental situation ([2], p. 55; [11], p. 194). Let us remember that the application of idealizational laws to reality can be carried out through a process of theoretical or experimental manipulation. The further refinement of the theory occurs insofar as the way in which the ideal conditions can be replaced by real conditions is specified each time in a more precise manner, and with this a better approximation of the law to the real conditions will be achieved. On the other hand, quantified idealizational laws make possible the refinement of the instruments, or of the manipulative technics of the experimental set-up, since they act as criteria to judge the reliability of these – for example, it can be judged that a given instrument which produces a vacuum is better than another because in the vacuum that the first one produces, the median velocity of two falling bodies is the same while the velocity that these bodies have when they fall in the vacuum produced by the second instrument, is not exactly the same.

Third, it is convenient that a theory have idealizational laws because these, given their greater simplicity (since they use few variables), clarify the theoretical field, and give us a better intellectual understanding of reality.

In Galilean terms, the use of idealizational laws clarifies the conceptual field because in these laws a distinction is made between essential or primary factors – such as the weight or gravity for the case of falling bodies – and accidental or secondary factors (although they may always be present) – such as the resistance of the medium (cf. [10], p. 26; in Galileo, [8], p. 74).[8]

Some authors claim that there is a strong relation between idealizational laws and theoretical foundations. For example: "it was in virtue of these idealized situations, that Galileo and Newton were able to make their great contribution to mechanics. Equally we should study the simple and ideal movements with great interest to obtain a real understanding of the foundations of dynamics" ([6], p. 342). I think that the fact that Galileo stated the law of free fall under the ideal condition of the non-resistance of the medium helped to establish the foundations of classical mechanics, insofar as in that law the true role that the resistance of the medium plays in the movement of the bodies is also established. What this law implies is that the resistance of the medium is not a necessary condition for motion to occur. This claim was fundamental for the emergence of a new mechanics, a non-Aristotelian mechanics.[9] Notice here that we have claimed that only some principles or theoretical foundations are idealizational. Not every principle is idealizational (for example, the Newtonian second principle, $f = ma$, is not idealizational because it is not stated to be valid under ideal conditions), nor is every idealizational law a principle (for example, the law of free fall, or the law of Boyle–Mariotte).

Therefore, idealizational laws have a basic theoretical function since they clarify the theoretical field by distinguishing between essential or primary factors, and accidental or secondary factors. Furthermore, some of them, precisely because they make this distinction, can become the foundations or principles of a theory.

Fourth, we see that scientific theories have requirements that are purely theoretical. One of these is that of systematization, or in the case of quantified theories, that of deducibility. Idealizational laws can be acceptable theoretically because they are deducible from other laws.

Galileo considered that his law of free fall (fall is uniformly accelerated) was theoretically acceptable, even if it was idealizational, because it could be proved theoretically (i.e., it was mathematically deducible from certain assumptions that were rationally acceptable, in particular, the assumption that the velocity is proportional to time) ([8], pp. 161–176). In other words, Galileo did not feel it necessary to look for another factual law of fall because his idealizational law satisfied the two basic requirements for the acceptability of a law: (a) it could be demonstrated theoretically, and (b) through experimentation (and/or a process of factualization) it fitted the empirical data.

Here we have tried to argue for the view that scientific theories are relatively independent of their relation with the sector of reality which they seek to understand; that is to say, the values that determine the acceptability of scientific laws include not only fitting the empirical data, but also others that are purely theoretical. For Galileo these values are those of mathematical demonstrability and simplicity ([8], p. 161). He considered these values to be so basic that, even when he could not specify in detail the process of factualization that would enable application, within an acceptable degree of approximation, of his law of free fall to the actual fall of real objects, he claimed that his law was acceptable because it satisfied the purely theoretical values of mathematical demonstrability and of simplicity.

Let us now turn to the problem of the evidence for idealizational laws.

We have already seen one kind of evidence that can be given for this type of laws: that they are theoretically demonstrable.

With respect to the empirical evidence, this can be at least of two types. (1) In many cases the empirical evidence can be direct, that is, it is possible to apply an idealizational law to reality – usually through experimentation. This occurs when the real or concrete value of the variables is small (that is, near zero), and for this reason the result that is obtained from the idealizational law strongly approximates the experimental (or experiential) result. This is the case, for example, for the law of free fall (that the fall is uniformly accelerated) ([10], p. 28). It is interesting to notice here that for some contemporary authors, a fundamental characteristic of idealizational laws is that they are *approximated* by some concrete phenomena or entities ([1], p. 266; Galileo sometimes seems to have this idea, cf. [8], pp. 253 and 256). (2) In other cases, the empirical evidence is rather indirect. The procedure consists in considering the idealizational law proven insofar as the real values that are obtained experimentally, when they are taken to a limit or extreme value (usually zero), give the same result that the idealizational law predicts ([1], p. 269; [10], p. 23). For example, Galileo proves that in the vacuum (ideal condition) all bodies fall with the same speed by arguing that in the case of very tenuous and poorly resistant media (real condition) bodies fall almost at the same speed, so that it is legitimate to suppose that if one gives to the real condition (i.e., a poorly resistant medium) an extreme value (i.e., a null resistance), bodies will fall with the same speed ([8], pp. 71–72). This kind of procedure is justified by the fact that the ideal conditions of idealizational laws are nothing but the idealization of real conditions, that is, they are real values taken to a limit ([10], p. 23).

II

In this section I will try to develop, in a schematic and tentative way, some ideas that are presupposed in the previous section. To this end, I will defend the following thesis: it was due to Galileo's work in physics that the use of *abstract theories* became legitimate in all (or almost all) scientific fields. Before him, given that the Aristotelian viewpoint was dominant, knowledge in the strict sense, or science, had to be developed through concrete or empirical theories. In very general terms, we can say that one of Galileo's most *revolutionary* contributions was precisely to show *how* an abstract theory of the sensible world could be developed. To defend this thesis I will consider first what an abstract theory is and why Galileo's theory is abstract; and second, why the Aristotelian theory is not abstract.

As is suggested in note 3 (cf. also notes 5 and 8), a concept, law or theory, is *abstract* if it does not correspond or does not fit in a *direct* way with *sensible* reality. As examples of abstract theories, we have the mechanistic physics of Galileo, Descartes and Newton. They are abstract because their object of study, matter with or without motion, and with or without forces that act at a distance, does not directly represent the sensible world: our perceptual world has dogs, stones, movements which are quick or slow, etc.[10] This idea is expressed by Dijksterhuis in the following way:

> Radically idealizing the phenomena by the elimination of all disturbing influences and schematizing everything with equal thoroughness by means of simplifying abstractions (e.g., by imagining all verticals to be parallel and by regarding bodies as material points), mechanics developed into an autonomous science, quite removed from physical reality ([5], p. 346).

There are at least two processes of abstraction that make possible the construction of abstract theories: mathematization (i.e. the introduction of mathematics into a non-formal theory) and idealization.[11]

It is precisely because the processes of abstraction, such as mathematics or idealization, do not yield empirical generalizations (cf. note 3), that abstract theories are *relatively autonomous or independent* of experience, that is, they can go beyond experience. Thus, for example, the fact that in Galileo's theory projectile motion is analyzed into two simple motions, the uniform horizontal and the accelerated vertical, shows that this theory goes beyond sensible experience. On the other hand, abstract theories are relatively independent of experience in the sense that, with them, principles or laws can be established on the basis of theoretical considerations, and not necessarily only on the basis of empirical considerations, that is, not every justification of a law or a principle consists only in its verification. Thus, we have seen that for Galileo the intratheoretical criteria for the acceptability of laws are those of mathematical demonstrability and simplicity.

Given our schematic characterization of abstract theories, it is clear that they have an *indirect* relation to the sensible world. In the case of mathematized theories, the use of measuring instruments will be required in order to be able to correlate the laws of such theories with the sensible world. Probably some process of factualization will be necessary (since, for example, the Earth is commonly considered to be a perfect circle touched only at one point by a tangent); and certainly, what is put in relation is not only a law but a set of laws, given the strong interrelation among laws that is implied by mathematical derivability. In the case where the theory is abstract because it is idealizational, its applicability to reality requires a process of factualization and/or the use of experiments. Indeed, what Galileo showed was how an abstract theory could be developed by demonstrating how such a theory could be applied to reality (principally through the process of factualization and the use of instruments in experiments).

Let us now compare abstract theories with concrete theories, such as the Aristotelian. The Aristotelian theory is concrete for the following reasons.

In the first place, the purpose of the Aristotelian physics is to apprehend theoretically or intellectually the sensible world, in all its complexity and diversity, in its concreteness. Thus, for example, the peripatetic law of motion takes into account all the factors that were considered to determine motion. On the other hand, there is talk of natural and violent motions, heaviness and levity, etc.

Second, its application to the sensible world is direct. The peripatetic law of motion tells us that the greater the force, or weight, the quicker the object will move, and the greater the resistance, the slower the motion will be; these are claims that we see confirmed in most of our everyday experience.

Third, the Aristotelian physical theory does not have idealizational elements nor is it mathematized. In particular, Aristotle held that physics could not be mathematized because mathematics is a theory in which change and qualities are abstracted or left aside, while physics purports to explain change in a world which is qualitatively diverse, that is, it purports to explain the changes that we perceive in the sensible world.

To conclude, thanks to the work of Galileo, physics changed from being a concrete theory to an abstract theory. This change was due to Galileo's use of idealization. Furthermore, because he used idealizational suppositions, Galileo was able to mathematize mechanics. And lastly, due to the introduction of idealizational laws, physics had to appeal to experimentation as a method to relate its abstract laws to empirical data.

NOTES

* The criticisms that Professors Cesáreo Morales, Mario Otero, Daniel Quesada and Margarita Valdés made to a previous version of this paper proved to be of great help in the elaboration of this version.

1 An example of a theoretical law which is not idealizational is the second law of Newton, $f = ma$. An example of a non-theoretical law (that is, a law which is not a principle of a theory) which is an idealizational law is the Galilean law of free fall. Therefore, the question of idealizationality is different from that of theoreticity.

2 This formula or law does not appear in Galileo. Presumably Dijksterhuis ([5], p. 336) and Koyré ([9], p. 241, note 102) formulated it on the basis of the Galilean claim that the speed of fall of a body in a resistant medium is determined by the excess of the specific gravity that the body has over the specific gravity of the medium ([8], p. 76).

This claim implies that $V = G - G_m$; the reason why $(G - G_m)$ must be divided by G is that if this is done then for the case of the vacuum $G_m = 0$, and with this we obtain $V = c(\frac{G}{G}) = C$, that is, we obtain the Galilean claim that in the vacuum all bodies fall with the same median speed or acceleration.

[3] The concept of abstraction is somewhat confusing. It is said that any generalization is abstract because it only takes into account common properties or aspects and it leaves aside those that are not common. On the other hand, it has also been thought that idealizational laws (which, like any law, are general) are abstract, although they are abstract in another sense, namely that they leave aside certain factors (or more precisely, they consider that their value is zero) which *are known to affect* the phenomenon referred to by such laws.

It is important to differentiated between abstraction (as a process or activity) and abstract (as a characteristic). I would claim that the process of abstraction can give rise to empirical generalizations or laws, but also to abstract or non-empirical laws. For example, it is usually considered that the empirical generalization 'All ravens are black' is reached through a process of abstraction in which only common properties are taken into consideration; but this generalization is not abstract because it directly describes something that is present as such in the sensible world. The idealizational law of free fall is abstract because it does not directly describe, or does not immediately represent, a phenomenon of the sensible world; in order to reach this law it was necessary to abstract, or leave aside, a factor that is known to affect the fall. Thus, the fact that idealizational laws require a process of abstraction does not imply that they are empirical generalizations or laws; indeed, they are abstract laws.

[4] Aristotle never stated this law in the mathematical form of a functional correlation given here. For Aristotle it was a qualitative statement of proportionality, and because of that somewhat vague ([9], p. 217).

[5] It is precisely because the abstract is obtained from the concrete (for example, the ideal circle that geometry studies is nothing but a concrete circle which has been perfected in the mind; or, the ideal conditions are only the real conditions taken to a limit and given the value of zero), why, in principle, it is possible to apply the abstract to the concrete.

[6] What Galileo literally says is that this method for calculating the real velocities follows from the principle that all bodies fall in a vacuum with the same speed ([8], p. 75). This deduction is not very clear, principally because Galileo does not explain how the weight of a body that falls in a vacuum intervenes, and for this reason it turns out to be somewhat peculiar that weight is introduced in the calculation of the real speeds. According to some contemporary commentators, the weight or specific gravity of a body does intervene in (indeed, it is the cause of) its fall in the vacuum; and the reason why bodies fall with the same speed in the vacuum, even though they are made up of different materials, that is even though they have different specific weights, is that the differences in weight (which depend upon the quantity of matter or mass) are counterbalanced by equal differences of mass (that is, of internal resistance to acceleration) ([4], p. 140; [9], p. 229). These authors realize that Galileo did not have the concept of mass; nevertheless, there is at least one passage in which Galileo seems to grasp this concept of mass or resistance to acceleration (cf. [7], pp. 213–215).

In general, Galileo's thesis about the influence of the resistant medium are not very clear. Sometimes he seems to confuse the resistance that the medium presents to its penetration by a falling body (mechanical action) with the upward thrust that the medium exerts upon the falling body (hydrostatic action). Thus, for example, Galileo claims that "The fact that any fluid medium diminishes the weight of a mass immersed in it, is due, Simplicio, to the resistance which this medium offers to its being opened up, driven aside, and finally lifted up" ([8], p. 81). On the other hand, he also speaks of the friction that is produced when a body moves through a medium; but his discussion is quite inconclusive (cf. [8], pp. 88–90).

7 An experimental proof of idealizational laws is possible because these laws more or less approximate some experiential data; to be more precise, this approximation occurs when, in real situations, the value of the variables that have been idealized is near zero.

8 The distinction between primary and secondary factors, or between factors that are taken into account by the law and factors that are left aside (that is, whose value is taken to be zero), is made on the basis of empirical and of non-empirical, or theoretical, reasons or only in terms of these last ones. Galileo's phrase, "the resisting medium is an 'external and accidental impediment'" ([7], p. 146) suggests that one of the reasons why he leaves aside the resisting medium in the law of fall is the philosophical idea (of 'Aristotelian' origin) that fall is a natural motion and because of that it is caused by the nature of the falling object, that is, by its weight, while the influence of the medium is external, and so accidental, and therefore secondary. In other words, the process of abstraction requires the use of some reasons that are not empirical.

9 For the Aristotelian mechanics, the medium is necessary for the occurrence of motion for two reasons: (a) The medium maintains the motion when the external motor that originated the motion does not accompany the moving object (a claim which depends upon the Aristotelian principle that all motion, both in its origin and in its continuance, must have a cause or motor that is in contact with the moving object) – for example, in the case of projectile motion. (b) The medium acts as a resistance which is necessary because, since $V = \frac{F}{R}$, if $R = 0$, then $V = \infty$, which is absurd for Aristotle. Against this Aristotelian conception of the medium, Galileo claims (1) that (a) and (b) are contradictory, since the medium cannot at the same time give motion and resist motion; (2) experience contradicts the claim that the medium bestows motion ([7], pp. 151–153); (3) experience contradicts the thesis that speed is proportional to the force or weight, and that speed is inversely proportional to the resistance or density of the medium; and (4) if the peripatetic law of motion is rejected, which must be done as the previous reasons indicate, then motion in a medium is possible (at least logically or theoretically) ([8], pp. 61-67).

10 It should be noticed that the question of the abstract character of a theory is different from the question of a realist interpretation of a theory. The mechanistic theories generally have been interpreted in a realist way by their authors (think for example of Descartes), that is to say, on the basis of such mechanistic physical theories they constructed an ontology according to which the real world consisted only of matter in motion or at rest (and with or without forces); and given this ontology it turns out that the physical theory directly described the real world. Nevertheless, even if we accept this realist interpretation, we can still claim that any mechanistic theory is abstract since it does not directly describe the *sensible* world, which is the world that should be explained, and also the world with which we interact. (It should be noted that the reason why a mechanistic theory is abstract has nothing to do with the fact that mechanistic

theories 'leave aside' the so-called secondary qualities, such as colors, textures, flavors, etc., since, programmatically at least, these theories try to *reduce* secondary qualities to primary (such as geometrical form, volume, motion, etc., that is, quantifiable properties), or, in other words, they try to explain the secondary qualities in terms of the primary ones.)

[11] It is convenient that we make explicit an idea that has been implicit throughout this paper. I consider as a plausible hypothesis the idea that some branches of mathematics in their origin were developed through processes of abstraction that are similar to those of idealization; that is to say, I think that some branches of mathematics have not always been formal sciences, constituted by analytical truths, but were empirical sciences – for example, Euclidean geometry was conceived to be the science of space. On the other hand, although many mathematical concepts (such as that of the circle, or of two, etc.), or some mathematical truths, may originally have been abstracted from sensible experience, this did not prevent the mathematical branches from later acquiring such a strong independence from sensible experience that they became autonomous in two senses: (1) the justification of their truths is totally intratheoretical, so that they become immune to empirical counterexamples; and (2) they can rise so far above empirical reality that they do not have any empirical applicability for a long time.

REFERENCES

1. Barr, William F. 'A syntactic and semantic analysis of idealizations in science'. *Philosophy of Science*, **38**, no. 2, 1971.
2. Barr, William F. 'A pragmatic analysis of idealizations in physics'. *Philosophy of Science*, **41**, no. 1, 1974.
3. Cohen, I. Bernard. *The Birth of a New Physics*. New York: Doubleday Anchor Books. 1960.
4. Crombie, A. C. *Historia de la ciencia: de San Agustín a Galileo*. Vol. II. Madrid: Alianza Editorial. 1974.
5. Dijksterhuis, E. J. *The Mechanization of the World Picture*. Oxford: Clarendon Press. 1961.
6. *Física*. Elaborated by the Physical Science Study Committee. Second edition. Vol. II. México: Editorial Reverté, S.A. 1966.
7. Galilei, Galileo. *Dialogue Concerning the Two Chief World Systems*. Translated, with revised notes, by Stillman Drake. Second edition. Berkeley: University of California Press. 1974.
8. Galilei, Galileo. *Dialogues Concerning Two New Sciences*. Translated by Henry Crew and Alfonso de Salvio. New York: Dover Publications. 1954.
9. Koyré, Alexandre. *Estudios de historia del pensamiento científico*. Second edition. México: Siglo XXI. 1978.
10. Krajewski, Wladyslaw. *Correspondence Principle and Growth of Science*. Dordrecht and Boston: D. Reidel Publishing Co. Series Episteme. 1977.
11. Kuhn, Thomas. 'The function of measurement in modern physical sciences'. In *The Essential Tension*. Chicago: University of Chicago Press. 1977.

12. Moody, Ernest. 'Galileo and his precursors'. In *Studies in Medieval Philosophy, Science and Logic*. Berkeley: University of California Press. 1975.
13. Nowak, Leszek. 'Galileo of the social sciences'. *Revolutionary World*. Grüner Publishing Co. 1974.
14. Shapere, Dudley. *Galileo: A Philosophical Study*. Chicago: University of Chicago Press. 1974.
15. Shapere, Dudley. 'Notes towards a post-positivistic interpretation of science'. In Achinstein, P. and S. F. Barker (eds.), *The Legacy of Logical Positivism*. Baltimore: The Johns Hopkins Press. 1969.
16. Weisheigl, J. A. *The Development of Physical Theory in the Middle Ages*. Michigan: University of Michigan Press. 1971.

RAFAEL MARTINEZ E.

AMONG MEN, CLOSEST TO GOD

No scientist, and certainly no other scholar in any field of thought has ever reached Isaac Newton's reputation. Only among legendary warlords, generals and great monarchs, can we find those deserving to occupy pedestals of equal standing. Born during Cromwell's revolution, Newton lived in an age of turmoil, spending his youth in the midst of the Restoration and a witness to the ascent to power of William of Orange: hard moments in which thoughts accelerate and become capable of understanding the new tides of the times and of answering the doubts and nonsense of the age. It was a time to dream new dreams and generate new ideas, to rebuild the world.

History, myth and legend have made Newton the genius of classical science and the architect of a new world vision. In this context adjectives acquire a well-defined meaning: they serve a strategy of opposition. To those delving in history the term 'classical' suggests that which took place long ago, before the big 'rupture', it is a reference to events preceding the instauration of modernity. In addition, it evokes a gesture of rejection to ancient science, quite close to common sense. There is another quality attached to the word 'classical': it denotes a style and therefore a culture. Thus, Newtonian science – or the Newtonian method of doing science – presents itself as a part of a set of human endeavours (economic, artistic, philosophical, theoretical and social practices) which together conform to what we call our civilization. The term also helps us to gaze beyond the appearances of autonomy conferred by the institutions and the academic practices of our times, recognizing in the web of western civilization the forms and shapes required by 17th century learning, when science and religion rested on metaphysical grounds which, instead of conflicting with each other, most often served as mutual complements. Newton is a classic, and for that reason his scientific style and the age in which he lived deserve a moment of meditation. It is not enough to perform the tasks of an historian; we must explore the different landscapes integrating his world, acknowledging the tensions erupting between science and culture – in a broad sense –

S. Ramirez and R. S. Cohen (eds.), Mexican Studies in the History and Philosophy of Science, 129–149.

and the subtleties hiding under the pageant of ruptures and changes of paradigms.

Science is not an abstract entity nor an impersonal construction. It is built by people in the same measure as any other transmitted knowledge, and is, in a way, an experience common to the human genus: designed by humans, it inherits their style. When a great man comes along, his legacy and deeds, therefore his style, inspire later generations. Three centuries have elapsed since the publication of the *Philosophiae naturalis principia mathematica* (1687) [1], better known as the *Principia*, and after having dominated the world of science up to the 19th century, the thought and writings of Isaac Newton are proof of the great sagacity and the solid disposition to look into the secrets of nature of a man who knew how to achieve the most ambitious synthesis of knowledge that had taken place since the times of Euclid and Aristotle. To many, the *Principia* is the greatest scientific enterprise ever produced by human mind. No text, except possibly Darwin's *On the Origin of Species*, has produced such notable changes in western thought. Understanding the breadth of Newton's achievements implies something more than just reading his works, many of them still unpublished: it requires knowing the background common to all thinkers of the 17th century and the range of Newton's legacy in contemporary thought.

FROM MAGIC TO SCIENCE

Many attempts have been made to date the 'renaissance' of what we might call the scientific spirit, a task that besides being arduous is also fruitless. There are so many elements to be taken into account that any opinion would face countless adverse considerations. However, we can easily agree that the structure of the Aristotelian universe began to suffer fractures at some point around the middle of the 16th century. The writings of those living during that period mirror the widespread belief on the need for changing things:

> I used to marvel and at the same time to grieve that so many excellent and superior arts and sciences from our most vigorous and antique past could now seem lacking and almost wholly lost. We know from remaining works and through references to them that they were once widespread. Painters, sculptors, architects, musicians, geometricians, theoreticians, seers and

> similar noble and amazing intellects are very rarely found today and there are few to praise them. Thus I believed, as many said, that Nature, the mistress of things, had grown old and tired. She no longer produced either geniuses or giants, which in her more youthful and more glorious days she had produced so marvelously and abundantly.

This was how Leon Battista Alberti expressed himself in the little book *On Painting* (*Della Pittura*, 1436) [2]. If by that year the notion of going through a Renaissance was something already accepted, then the idea of an almost total disappearance of artistic, mathematical, and rhetorical genius was just plain nonsense. But anyhow, the myth postulated by Alberti had a certain power, persisting at least as a rhetorical flourish and supported by the ancient Greek and Roman cultures. In the High Middle Ages there was an obsessive preoccupation with a vanished 'Golden Age' of sages in possession of all the knowledge that was then being laboriously rediscovered. What was new in the Renaissance myth was the flagrant ignoring of the glories of Gothic architecture and sculpture, of *Trecento* painting, and the not so old works of Dante and Petrarch [3]. On the other hand, though a much greater degree of continuity with the Middle Ages had been revealed, it was without question that something had happened to make the world of Machiavelli and Leonardo da Vinci a very different place from that of St. Thomas Aquinas. In fact, Alberti himself summons all those whom as ... [2]: "Fillippo, Donato, Luca and Masaccio, possess genius for accomplishing every praiseworthy thing"

> Therefore I believe the power of acquiring wide fame in any art or science lies in our industry and diligence more than in the gifts of Nature. It must be admitted that it was less difficult for the Ancients. Our fame ought to be much greater, then, if we discover unheard-of and never-before-seen arts and sciences, without teachers or without any model whatsoever.

Alberti is in effect claiming that it is not only a Renaissance of a past glory, but a 'naissance' of new visions and achievements. It is important to recall that the "new arts and sciences" to which Alberti lays claim, or its equivalent in other languages, did not normally acquire its present meaning until the 17th century, and by no means consistently until the 19th [4]. The discovery and dissemination of old pieces of Greek literature and the widespread knowledge of Greek in Italy had a profound effect on the pattern of university teaching: the *Trivium* (grammar, rhetoric and philosophy) gave more emphasis to literature, history and moral philosophy, to the detriment of logic and 'physics' [5].

The fundamental basis of the Faculty of Arts was constituted by the *studia humanitatis*. In the same Faculty there would also have been taught at least some of the *Quadrivium* (arithmetics, geometry, music and astronomy), these studies being regarded as approximating most closely to the true nature of 'science'. The distinction between 'science' and 'art' had no necessary reference to subject-matter but only to *certainty*, that is, knowledge that follows necessarily from unquestioned premises. Under this, optics, geometry, and arithmetic (theory of numbers, to be distinguished from mere calculation, then called 'algorism' or 'logistic') were as near to certainty as any secular knowledge could be. Astronomy and music (theory of harmony), being the mathematical representations of celestial phenomena, could be accepted as 'sciences' without much argument. 'Physics' was not a science but, as its name implies, the study of nature – the realm of coming to be and passing away – sharply distinguished from the changeless heavens, which are, have been, and will be equal to themselves, comprising everything including and beyond the sphere of the Moon. This, the radical distinction between 'nature' and the 'heavens' was one of the most serious obstacles to the emergence of modern science [6]. The other obstacle was the absence of the concept of natural law [7].

Fifteenth century scholarship shows a complex intertwining of discourses coming from disciplines as diverse as magic, alchemy, mysticism, empiricism and natural history. For some historians, such as Frances Yates [8] and D. P. Walker [9], the magician of the Renaissance is the immediate predecessor of the 17th century scientist, and its conceptual framework is the condition which fostered the appearance of the new science. Galileo's science, which no one would deny was a participant in this process of building the new premises of science, can be characterized by the use of mathematics in physics and for its appeal to experiments. This new science disdained traditional schools of thought and found support in its preoccupations with technical problems of engineering, nautical and military issues, and in the use of special materials and instruments of relative complexity. According to Ms. Yates, the same qualities can be found in the hermetic philosophy of the Renaissance, a new vogue of the times mingling neoplatonism, magic, numerology, kabbalah and alchemy. In this mixture it is possible to identify two converging pathways: one of them dealing with neoplatonic magic, a kind of magic derived from the old belief in Emanations coming from the original 'one'. The hermetic mystic thought himself capable of controlling these 'energies', of exploiting the 'sympathies' that held the Cosmos together. Alchemy and astrology gave hints as to how to proceed and how to concentrate power in amulets and symbols. In the act of performing these tasks – the gathering of magic elements, charms, burning mirrors, harps, etc. – the hermeticist became the magician and in a certain manner the ancestor of the experimental physicist. The second pathway fuses Pythagoreanism and the kabbalah. The marriage between the two systems of beliefs is achieved by finding Number among the first Emanations, ideas springing from the Great Mind, and calling forth the magic of numbers [8]. This line of thought leads us to Galileo in two ways: providing support to an unrestricted use of mathematics, and authorizing the natural philosopher to use the simple mathematical relationships existing between the physical quantities which constitute the 'Pattern of the Universe'.

From Ficino to Galileo we can hardly find a thinker who does not touch upon some of the hermetic subjects. However, this does not mean they were magicians; it only reflects the fact that magical thinking permeated many aspects of life in the Renaissance: mechanical designs, filters and cosmetics, cryptograms, mirror games, magical lamps, hymn singing and the invocation of angels or demons. All of them were part of natural procedures aiming at controlling or mastering events of everyday life.

Religious thought and superstition went hand in hand, and even though the first attempted the suppression of the latter, on occasion religious festivities and rites bordered dangerously on the limits of that which it attacked. In spite of that, experience began to yield its harvest, and little by little doubt began to fall upon the belief that people possessed power over nature. Thus began the process through which mankind learned how to manipulate nature by means of its own rules. White magic, falling into discredit, gave way to natural law [7]. The process was slow, with many traps and false sidepaths. Francis Bacon himself – held by many to be the father of experimental science – resorted once in a while to supernatural considerations blended by the ideological pressures of the times. His attempt to develop a mathematical astrology in which witches and demons played no part, shows that Bacon still believed in a stellar determinism which ruled over all matter, including the human body and its appetites. The human soul was unique in avoiding the ruling of the stars, thus reconciling his position with that of St. Thomas Aquinas. Man is free from restrictions . . . "man can do what he wills, he is a *magus*, a miracle; he operates like a god, he is a *Magnum Miraculum*, deserving reverences and honours". These words reverberate deeply in the writings of Pico della Mirandola, who in the *Oration on the Dignity of Man* [10] situates humanity at the top of everything that was created by God. From that position Man only has to choose the best path to follow.

THE END OF ARISTOTELIAN SCIENCE

At the beginning of the 16th century the revival of Platonic thought was replacing the old Aristotelian machinery that had supported to the beliefs, doctrines, governments and hierarchies characterizing the Middle Ages. Changes in ways of thinking not only originated in universities and their surroundings; things were happening around the world and some of them, like the discovery of America and the invention of printing, were not immediate consequences of a new science, but the result of the emerging 'technologies'. Problems such as the government of colonies under different climatic and cultural conditions, together with the new experiences arising from changes and improvements, demanded radical adjustments in the relatively hierarchical system at work throughout Europe. It is hard to say to what extent these facts influenced scientific method, even if we can consider them as a product of the *Zeitgeist*, a strategy that even if it does not explain at least emphasizes belief in a common origin of most of the changes taking place. Be that as it may, it seems acceptable to say that gradual replacement of 'rational' theories supporting the state of affairs by theories with an empirical bias reveal a new world of thoughts.

Copernicus's great intellectual and posthumous achievement in cosmography was heralded by events, declarations and tragedies that paved the way: before Copernicus, Galileo or Kepler, it was Giordano Bruno who reduced the medieval cosmos to ashes by abolishing the barrier separating terrestrial and supra-lunar realms. This cosmical ruin of the Aristotelian world must have played the leading part in John Donne's lament: "New Philosophy calls all in doubt . . . 'Tis all in pieces, all coherence gone, all just supply and all relation".

It is a hard task to say when the Renaissance ends: Bacon, the 'Great Instaurator', never really grasped the character of the machinery of actual scientific discovery. Those who did – Galileo, Gilbert, Kepler, Harvey – were in many ways linked to the past: their conclusions form the basis of much of modern science, but their methods and concepts were still highly coloured by the myths of the Renaissance [11].

Finally, the conceptual tensions that had been gathering reached a climax and gave birth to the first attempts to build a new philosophical discourse, better suited for a universe which was no longer burdened by the difference between the sacred and the profane, and no longer accepting a perfect and everlasting place above the Moon and an imperfect and corruptible world here on Earth. First came Nicholas of Cusa (Cusanus) and his new idea of infinity [12], followed by Copernicus and Kepler, accouchers of the system which was about to displace the Ptolemaic–Aristotelian world. This new system portrayed an alternative way to read the universe, a task to which Galileo applied himself. In assuming the role of apologist for the new scientific method, Galileo drew out the venom of entrenched, institutional resistance to a scientific revolution [13], and freed those who would follow him from fighting the same duels: they could assume the task of developing the newly born method. Their names, Descartes, Newton and Leibniz, are known by anybody with the most casual acquaintance with modern thought, and their works can be summarized by saying that Descartes was destined to provide the philosophical foundations of the new method, and Newton and Leibniz, working with the Cartesian edifice, made dramatic reforms that led to the mechanistic vision of the world [14].

Much of pre-Cartesian thought was devoted to justifying the proposition that the mathematical core of the method, when used to study natural phenomena, can reveal the Universe's innermost reality. This obsession can be seen in Galileo when he gives great emphasis to the capacity of the mathematical essence of the new method to reveal the underlying reality of nature [15]. The second point of concern for Galileo is the singleness of the mathematical postulate system. For Descartes and Newton the primary issue is the capacity of scientific method to be the paradigm of knowing, the best available approximation to a universal language, a language longingly sought-after by so many medieval thinkers. Their lack of interest in the method's true capacity for apprehending the innermost nature of things seems to be the reason why they were able to develop the method's potentiality to answer the question:

how can man acquire true knowledge in an infinite universe? Not being constrained to defend the connection between the mathematical essence of the method and the nature of physical reality, they achieve a level of thought that goes beyond the celebration of the infinite universe, and because of that, they are able to examine analytically the foundations of this new kind of knowledge [16].

THE METAPHYSICAL FOUNDATION OF MODERN SCIENCE

If the title of Prophets of the New Universe can be given to Copernicus, Kepler, Galileo and Bruno, that of Father of Modern Philosophy belongs to Descartes. Kepler, while keeping Aristotelian philosophy away from his work, was still unable to replace it with another philosophical foundation. Galileo, unable to show how his methods adapted to deeper processes of thought, failed in his attempt to show his contemporaries how to think. Bruno's enterprise collapsed when he could not offer anything besides the claim of the supremacy of the intellect over the absurd ruling of the authorities. All these supreme scholars, even while acting as priests during the birth of the new science, fell short of articulating a firm philosophical basis for the method(s) they were using.

Descartes could criticize Galileo for assuming that it was valid to work without a metaphysical foundation. His method did not delve into the 'first causes of nature' and was restricted to explaining only 'certain particular aspects', thus constructing 'without a basis'. Descartes immediately grasped what was at stake: a new science could not survive without a metaphysics. In the same manner as men need a rational basis ensuring their salvation, they also reach for a primary intellectual support for the new science.

Descartes' endeavour reached vast proportions. He travelled to many places, collecting experiences and studying the great books of the world, until one day he decided to become the object of his own scrutiny, starting his search for truth by doubting everything learned up to that moment. His metaphysical mission started with a mystical experience which took place during the day and night of November 10, 1619. In the solitude of his chamber he began a skeptical revision of the knowledge he had attained; and he found that uncertainty and lack of unity seemed to characterize this knowledge both in the arts and in science. In mathematics he found certainty, but this certainty failed to shed clear light on other fields of knowledge. What was needed was a fresh start, one that would sweep away all systems which had become encumbered with false logics and half-truths. Gradually an idea grew in his mind that would allow him to reach the certainty of mathematics in other fields which, like philosophy, appeared in a state of chaos. After many hours of taxing his mind to its utmost, he found the desired method: as in a blinding flash, he saw the unity of all the sciences, of all knowledge [17].

Exhausted, when the day had reached its end, he retired to sleep and had three dreams, which became the origin of his *Meditations on First Philosophy* (1641). In these works we find the three principles of his metaphysics [18]: (1) intuition of the necessity of the thinker's existence; (2) whatever is true is clearly and distinctly perceived as an idea, and whatever is clearly and distinctly perceived as idea is true; (3) the existence of a veracious God. Capable of conceiving God as a truly infinite being, but incapable of conceiving Him as non-existent or imperfect, as lacking in veracity, Descartes succeeded in vanishing all the doubts that troubled him. The Cartesian metaphysical system was thus completed, and the three first principles might be said to find their expression in the formula *Cogito ergo sum, ergo Deus est.*

A question was still pending: how can these principles be used as basis for claiming scientific certitude when investigating the physical world? The answer is given in the *Rules for the Direction of the Mind* (written about 1629) [17]. There, Descartes establishes that scientific method is nothing more or less than pure thought confronting the conditions of its own truth. Mathematics is the discipline that most obviously exemplifies this. Science uses what is clearly exemplified in mathematics in order to harmonize pure thought and the structure of the physical world. Even more, science not only replicates mathematical language but also brings the certitude exemplified in mathematics to knowledge of the natural world, in such a way that it transcends any dependence on empirical observations. Science becomes, then, pure methodology – knowledge knowing itself in its purity – and this ensures that the method used to penetrate the natural world is operational.

Whenever one thinks about an object, this object exists even if it is not in the mind of anyone. Cartesian metaphysics can never *directly* establish the existence of bodies since it can establish no existence independent of pure thought – with the exception of the finite thinking self and the perfect infinite substance – but it can indirectly offer support for the claim of science to be dealing with physical bodies that have an existence independent of thought. This evidence takes root in the certitude of the existence of a veracious God, who by His own nature would not allow science – being purely necessary thought – to be deceived in its most important claim, the existence of an external world susceptible of being apprehended by human beings. Thus, between the "mathematical domain of essences and the corporeal portions of Nature's substantiality there obtains a correlation. It is there by divine institution" [18].

The next step in the Cartesian project was to develop the method which would provide scientific knowledge of the world. So brilliant and clear was its structure that many of his contemporaries and many of his followers were more fascinated by his contributions to the method than they were by its metaphysical foundation. Essentially two broad divisions were considered: the first starts with given data and proceeds to separate them into their components, looking for significations; this is the division known as *analysis*. Second, it then goes on to build up knowledge that is not only certain but genuinely new, knowledge that was not contained in the original data. This is the division known as *synthesis*. The end product of the whole method is the merging, in a single articulated vision, of the analytic ground work and the synthetic superstructure, thus reaching a sacramental experience of the world as a single limitless fabric, infinite in its extent and novelty. In this manner the Cartesian method acquired the certainty of its ability to know the totality of that which falls under its scrutiny. The thinking subject, being the one who craves after knowledge, is restricted by his own finite human mind to penetrating only the essence of that with which it can establish a relationship. The result is that physical things *qua* physical things, and natural essences as natural, are alienated from finite thought in its purity.

Human thought is fated to deal with nature not as nature, but with nature as thought, capable then of revealing only the essence of finite thought and not that of physical objects. This situation places before us a 'terminal dualism' which Descartes cannot avoid, but instead he uses it in order to establish a new program for science. The dualism bequeathed by Descartes calls forth a new dualism – between mind and matter – to replace the classical dualism of ancient thought between earthly and celestial entities, the imperfect *down* and the perfect *up*. Later on the Cartesian dualism will develop into our modern dualism between what happens inside human consciousness and what takes place on its outside. This was how the quest for certainty started by Descartes leads us to a strange situation: an infinite universe which in its essence avoids the grasp of the human mind [18, 19].

THE MECHANICAL UNIVERSE

Two ideas played a fundamental role in the development of science: order and cause. Concepts attached to them had a great influence on the paths followed by the creators of modern science, providing new answers to questions which were not answered satisfactorily by arguments given before. To the question, why does a stone fall when let loose?, or, to get nearer to the person who gave the correct answer, why does the apple fall to the ground when leaving the tree?, the Middle Ages responded, under the light of Aristotle, that it goes downward and not upward because its nature is to follow the downward direction. What is important in this is not the apparent tautology but the fact of confronting a generalization: it is in the nature of all apples to move downward. Even if this seems to be trivial it is not so, since the capacity to generalize a concept is something characteristically and fundamentally associated with intelligence: to uncover sameness among non-identical objects, and, when necessary, to perceive identities among elements apparently not related.

The well-known newtonian anecdote about the falling apple illustrates this point: Newton was able to perceive the sameness, until then never perceived, between the fall of the apple and the orbiting of the Moon around the Earth.

As already pointed out, establishing resemblances became a more fruitful endeavour, especially when it led to abolishing the differences between celestial objects and terrestrial matter. Medieval conceptions of matter made use of notions quite foreign to our modern way of understanding. It was not unusual to believe in causes possessing something similar to the human will. Earth, air, fire and water had natures ultimately akin to our human mind, and were driven by animal will: bodies fall because they want to reach the state of rest at the center of the Earth, fire wishes to go upward ... and everything taking place is reduced in the last analysis to these volitions. This in turn presupposes an order toward which Nature's 'project' converges. Its aim is to reach the status of the grand ideal hierarchy, the one ruling over everything created, perfection incarnate. This line of argument was abandoned because it was considered to be exhausted; it did not explain known facts, it lacked coherence, and on occasions made no sense. By the end of the 17th century this world vision was replaced by a new concept: a Universe whose functioning resembles a blind machine where automatically everything happens simply because something happened before. The world became a machine producing events, a view known as the mechanical philosophy [20].

The title of Father of the Mechanical Philosophy cannot be ascribed to a single person. It is clear that during the first half of the 17th century, scientific circles in Western Europe had already accepted the mechanical conception of nature which, with the support of Kepler and Galileo, reached maturity in the writings of men like Mersenne, Gassendi and Hobbes. However, it was Descartes who gave the greatest leverage to this new vision, providing his proposals with a philosophical *rigueur* that exceeded the achievements of any previous thinker. The mechanical philosophy of nature has as its basic tenet that the world is a machine whose components are inanimate bodies, that it undergoes changes caused by physical necessity and that it remains indifferent to the existence of thinking beings. As an immediate consequence, the world of sensory experience was reduced to mere illusion.

One of the touchstones of the new philosophy was the principle of inertia. Taking the premise that 'all natural phenomena are produced by particles in motion' leads to questions concerning the very cause of motion. Matter, by definition being inert and lacking any active principle, does not qualify as the cause of its own movements. During the 17th century it was taken for granted that the origin of motion could be traced to God. At the beginning He created matter and put it in motion. What keeps that motion going? The only acceptable answer, giving the total rejection of active principles, was the principle of inertia: nothing is required in order to keep matter in motion. Motion turns out to be a state of matter, and like any other state, it remains without change as long as there are no external agents to alter it. Such an external agent could be an impact, causing a transfer of motion, but never its destruction.

Descartes' universe was a *plenum* and by identifying matter with extension he concluded that space must, by definition, be totally filled with matter, or even better, must be matter. There is no void, and because of this, motion is possible only when bodies change their position to occupy a place that was simultaneously emptied, giving birth to a motion which forms a closed circuit. This 'circular' motion – a vortex – generates centrifugal pressures in the plenum that turn out to be the cause of the most important natural phenomena. The Cartesian image of the vortex containing our solar system is a whirlpool so big that, in comparison, Saturn's orbit represents nothing more than a point. Putting together some other considerations, Descartes finally developed a system which in a plausible way could replace the Aristotelian picture of the concentric spheres [14].

It was in this fashion that the mechanical philosophy gradually defined the area of concern of the 17th century scientist. His language became the condition of questions posed and answers given. So rudimentary was their machinery that some other fields of knowledge were unable to advance further due to the absence of an appropriate language. To a certain extent, mechanical discourse became a barrier to the Pythagorean conviction of a Nature capable of being described in mathematical terms. The tension between both currents of thought was finally solved by Isaac Newton.

NEWTON'S SYNTHESIS

The 17th century, justly known as the century of genius, could boast of possessing so many brilliant minds as to outshine the firmament itself with all its stars. And the brightest of them all were Descartes and Newton. According to Koyré, the first one "conceived the ideal of modern science or its dream" – the *somnium de reductione scientiae ad geometriam* – and the second one "firmly put physics back on its own feet" [21].

In order to understand the history of modern philosophy, its paths and compromises, we must turn to the major innovators of scientific thought, the greatest of which was Isaac Newton. One century after Newton's age, nearly all philosophizing, especially in its more technical aspects, was an attempt to come to terms with, to analyze and criticize, and to extend his thought. Although Newton never reaches the heights of Descartes' metaphysical synthesis, his methodological synthesis goes beyond anything done before by transforming previous discoveries into a coherent whole.

The magnificence of the Newtonian synthesis resides largely in a welding together of the two primary strands of 17th century thought: the mathematical rationalism of the continental tradition and the physico-mathematical experimental learning cultivated by the Royal Society in England. In the manner of Galileo, but on a much grander scale (because of his access to a much more mature and precisely formulated methodology, for he walked in the path paved by Descartes), Newton managed to effect his synthesis of the mathematico-deductive and experimental dimensions of scientific method. As one of Newton's commentators remarks [22]:

> Newton's laws of motion were primarily conceptual in origin and ... were probably derived directly from Cartesian physics. It cannot be claimed that these laws were originally supported or derived inductively from detailed experimental work ... this lack of empirical support ... was not an accidental feature, for no amount of experimental work could ever establish a principle of natural motion ... this is a consequence of using the concept of force ... which exists only to explain effects. Since we are at liberty to choose to regard whatever states we please as natural ... it follows that the forces which we should say are operative in nature depend upon the choices we make. For this reason, then, a principle of natural motion ... cannot be an empirical principle.

If the interpretation is correct, then no experiment nor empirical observation can guarantee the validity of the principles of the Newtonian system of nature. The only guarantee, as suggested by Koyré, is the fact that Newtonian experiments "presuppose in an axiomatic fashion a mathematical structure of nature". If we remember the often quoted phrase *Hypotheses non fingo* (I do not feign hypotheses) [1], Newton seems to say that he works with forces which are only mathematical in character; and if it were not so he is dealing with them only insofar as they are subject to mathematical treatment.

But the coincidence between the philosophical principles of Newton and Descartes was not perfect. There are two ways in which Newton undertook a revision of the Cartesian enterprise in modern science: the first regards the Cartesian method itself, and the second deals with the revision of the relation of metaphysical issues to scientific method and the acquisition of knowledge.

Newton's first objection concerns the capacity of the mathematical deduction to supply, in and of itself, a logical proof sufficient to establish a scientific conclusion as factual. This opinion is illustrated in the following passage from the *Principia* [1]:

> To describe right lines and circles are problems, but not geometrical problems. The solution of these problems is required from mechanics, and by geometry the use of them, when so solved, is shown.

The deductive character of the scientific method establishes some measure of certainty, but the full solution to its problem is not reached until this certainty is shown to be applicable to natural phenomena. The applicability to natural phenomena is demonstrated by the confirmation lent by experiments, and this verification is the prime agent for transforming certain knowledge into *real* knowledge. As Newton himself remarks in the *Opticks* (1704): "My design in this book is not to explain the properties of light by hypotheses but to propose and prove them by *Reason and Experiment*" [23].

Summing up what has already been said: the Newtonian revision of Descartes' method consists in expanding the method of analysis and synthesis by the inclusion of a third element: experimental verification.

The second major Newtonian revision of the Cartesian enterprise refers to the relation of metaphysical issues to scientific method. Newton appears concerned to keep separate the issues of method and the issues of metaphysics and religion. Being a devoutly religious man, and in many ways a mystic, he did not believe that religion or mystery had anything to do with the mathematical laws of planetary motion, although they had to do with the First Cause.

Contrary to Descartes, Newton saw method as presupposing no metaphysics for its exercise, even though at some point this exercise leads to certain metaphysical and religious realizations. The situation becomes clear if we recall that Descartes approaches God's existence from the viewpoint of the isolated thinking subject, whereas Newton approaches his existence in a more traditional (almost Scholastic) fashion from the order of nature, thus seeing God as the First Cause. This does not mean that God's existence is any less important for Newton than it is for Descartes, it is simply that existence functions in a different way in their syntheses: in Descartes as the ground of scientific method, in Newton as the outcome of the exercise of that method. In restoring the notion of First Cause to nature, Newton reveals a deep intuition for the limits of the purely mechanical interpretation of nature. The book of nature may be a very complicated mechanism, but still it requires the active interference of God to be sustained [24]:

> God ... who not only had made the world clock, but who continuously had to supervise and tend it in order to mend its mechanism when needed ... thus manifesting his active presence and interest in his creation.

Newton, worthy of occupying a place such that "Nearer the gods no mortal may approach" (*Isaac Newton, an Ode*) [25], was unable to overcome the Cartesian alienation between the thinking being and the very essence of things. Newton himself was quite explicit on this matter when he declared that scientific method was inadequate to penetrate an object's innermost nature. In a letter to Bentley – one of his commentators – concerning this issue Newton says: "you sometimes speak of gravity as essential and inherent to matter ... pray do not ascribe this notion to me, for the cause of gravity is what I do not pretend to know ..." [26].

To Newton, gravity was the key to his *systema mundi*, and in spite of that he took for granted the impossibility of ever grasping its essence, aspiring only to reach a complete knowledge of its behaviour. Even in accepting this limitation science is not diminished, since even then it does provide a perfect description of nature's behaviour, clearly deducible starting from human experience and always verifiable through experiments [27].

To posterity the Newtonian synthesis became the focus of convergence of many important philosophers. Some aspects of Newton's thought were followed by John Locke and reached their peak in David Hume's natural empiricism. Other aspects were taken up by Euler and through many intricacies became part of the grandeur of Kant's critical philosophy, probably the most profound analysis of Newtonian mechanics.

THE MYTH

Galileo died the same year Newton was born, a simple coincidence suggesting the idea of an heir. His infancy in no way provides any clue about the future of someone who would live to 84 years – alone, arrogant and irascible – and would become one of the most perceptive minds that ever lived.

Praises for Newton can hardly be qualified as exaggerations. New findings pertaining to the chronology and the way in which he achieved his discoveries and systematizations show us an intellect breaking through the barriers of his time. Like Shakespeare when using Romeo's lips to say "... it seems she hangs upon the cheek of night/Like a rich jewel in an Ethiop's ear...", Newton established deep similarities where no one had seen them: between the majestic movement of the Moon and the no less elegant of the apple when falling to his feet.

In the year 1666, in a series of astonishing perceptions, each one itself worthy to ensure him eternal fame, Newton conceived the idea that gave birth to modern science, a science built to suit man, the measure of all things.

In spite of the long and important periods of intellectual maturation, we must not undervalue the moment of perception, the quasi-divine experience in which new horizons suddenly jump forward and the old rules and regularities characteristic of orthodox thinking appear to vanish. The act of creation overflows imposed limits, showing new connections, and attempting to weave its own confinement, a symbolic act originated in the longing for a totality which has been finally and wholly embraced. New images, new miracles of the imagination will once again open the system, revealing similarities among everything which obeys the new laws. A highly unlikely communion, but one that finally gave exceptionable gains to Newton, thus came to be the act of creation. Newton took facts, methods, systems and superstitions that pretended to explain our universe, and looked beyond the superfluous, to hold what was essential and to build whatever was necessary in order to become the architect of the most impressive explanatory system for our physical universe that the world had ever known.

Much has been written about Newton; in particular his complex personality has attracted the attention of historians and scientists throughout history, and because of that, multiple controversies have arisen. One of the most interesting has its origins in the following account due to Newton's own hand [26]:

> In the beginning of the year 1665 I found the Method for approximating series & the Rule for reducing any dignity of any Binomial into such a series. The same year in May I found the method of tangents ..., & in November had the direct method of fluxions & the next year in January had the theory of Colours & in May following I had entrance into ye inverse method of fluxions. And the same year I began to think

> of gravity extending to ye orb to the Moon & (having found out how to estimate the force with wch globe revolving within a sphere presses the surface of the sphere) from Keplers rule ... I deduced that the forces wch keep the Planets in their Orbs must [be] reciprocally as the squares of their distances from the centers about wch they revolve: and thereby compared the force requisite to keep the Moon in her Orb with the force of gravity at the surface of the earth, & found them answer pretty nearly. All this was in the two plague years of 1665 & 1666. For in those days I was in the prime of my age for invention & minded Mathematicks & Philosophy more than at any time since.

This text, quoted so many times, provides the basis for the myth of Newton's years of wonder, his *anni mirabiles*. Free from academic pressures, Newton opened the gates for a rush of new ideas and in a miraculous burst of creativity he made the three discoveries on which his eternal fame stands.

Recent studies call attention to inaccuracies in the accepted chronology of events. It has been established in a reliable manner that even if we take for granted that Newton started working around 1666 on the subjects he mentions in the above passage, the actual results that can be referred to that year differ considerably from what he published years after. Then Newton still had to undergo a period of maturation and acquisition of knowledge that finally led to the *Principia* (1687) and the *Opticks* (1704). In all his contributions, it is possible to see and to analyze the influence of other great thinkers of the past: Galileo and Kepler inspired his search for physical causes for natural phenomena; from Descartes he took optics and mathematics. Taking the achievements of the french thinker as a point of departure, Newton was in a superb position to embark upon a more ambitious synthesis of knowledge. From Wallis and Barrow, his teachers at Cambridge, Newton took the tools to build the method of fluxions, whose later evolution made what we now call the differential and integral calculus. In Hooke and Huygens he found a way to describe circular orbits which proved fruitful for the postulation of the law of universal gravitation.

These comments delineate a more complete and precise image of the man of genius. Close scrutiny of Newton's extant material still waiting for publication [28], reveal a life that cannot be summed up in terms of blinding flashes of insight, one following the other. What we discover is a man performing incredible amounts of systematic work – Newton had the astonishing capacity to concentrate on a problem and spend endless hours without eating or sleeping – which begins by building upon the prior work of others and continues on the basis of his own successes, large and small. The manuscripts show that he made mistakes and learned from them, that on occasions he took false turns, and that not always was he able to comprehend immediately the full implications of his own ideas. The man we confront seems more human, and all the more deserving praise.

Newton's interest covered other grounds which have not attracted much interest in the past. It is only recently that scholars turned their eyes to the alchemical and theological works of Newton [29–31]; without them the image of the great scientist remains incomplete. But in spite of that, one can be certain that the scrutiny of these non-orthodox subjects will not alter the admiration for the one who was able, with one stroke – the concept of universal gravitation – to reveal the nature of Kepler's three laws of planetary movement, to refute the Cartesian vortex theory, to solve the riddle posed by the origin of the tides, and to explain the Galilean claim that all bodies fall to the ground with the same acceleration.

Newton voyaged over many strange seas of thought, voyages from which more than one adventurer of the seventeenth century failed to return, and he returned carrying coffers filled with treasures. During his lifetime he attained everlasting fame and on his death, according to Voltaire, he received homages "worthy of kings". Allowing our mind to drift back to those days we can hardly avoid thinking about the old saying "The King is dead" and to add that, on this occasion, there was no heir.

> To refuse and to yield ... how delightful; to pursue and conquer, how august; to perceive and to reason, how sublime.
>
> (*Orlando*, Virginia Woolf)

REFERENCES

1. Isaac Newton. *Mathematical Principles of Natural Philosophy and the System of the World.* A revision of Andrew Motte's translation by Florian Cajori. University of California Press. Berkeley, 1982.

2. Leon Battista Alberti. *On Painting*. Translation by J. R. Spencer. Yale University Press. New Haven and London, 1966.
3. J. A. Mazzeo. *Renaissance and Revolution*. Methuen & Co. Ltd. London, 1967.
4. E. García Estébanez. *El Renacimiento: Humanismo y Sociedad*. Ed. Cincel. Madrid, 1986.
5. E. Garin, *Educazione umanistica in Italia*. Editori Laterza. Roma-Bari, 1975.
6. W. Wightman. *Science in a Renaissance Society*. Hutchinson and Co. London, 1972.
7. J. Bronowski. *Magic, Science and Civilization*. Columbia University Press. New York, 1978.
8. F. A. Yates. *Giordano Bruno and the Hermetic Tradition*. Vintage Books. London, 1969.
9. D. P. Walker. *Spiritual and Demonic Magic*. University of Notre Dame Press. London, 1975.
10. Pico della Mirandola. *Oración sobre la Dignidad del Hombre*. Luis Martínez Gómez (Ed.). Editorial Nacional. Madrid, 1984.
11. H. Butterfield. *Los Orígenes de la Ciencia Moderna*. Taurus. España, 1982.
12. A. Koyré. *From the Closed World to the Infinite Universe*. The Johns Hopkins University Press. Baltimore and London, 1979.
13. R. A. Hall. 'The significance of Galileo's thought for the history of science'. In *Galileo, Man of Science*, E. McMullin (Ed.). The Scholar's Bookshelf. Princeton, 1988.
14. E. J. Aiton. *The Vortex Theory of Planetary Motion*. Elsevier. New York, 1972.
15. S. Drake. *Galileo at Work*. The University of Chicago Press. Chicago, 1978.
16. A. Balz. *Descartes and the Modern Kind*. Archon Books. New York, 1967.
17. R. Descartes. *Reglas para la Dirección del Espíritu*. Translation by J. M. Navarro. Alianza Editorial. Madrid, 1984.
18. L. J. Beck. *The Metaphysics of Descartes: A Study of the Meditations*. Clarendon Press. Oxford, 1965.
19. R. Descartes. *Discurso del Método*. Translation by R. Frondizi. Alianza Editorial. Madrid, 1983.
20. E. J. Dijksterhuis. *The Mechanization of the World Picture*. Princeton University Press. New Jersey, 1986.
21. A. Koyré. *Newtonian Studies*. The University of Chicago Press. Chicago, 1965.
22. B. Ellis. 'The origin and nature of Newton's laws of motion'. In *Beyond the Edge of Certainty: Essays in Contemporary Science and Philosophy*, R. G. Colodny (Ed.). Prentice Hall. New Jersey, 1965.
23. Isaac Newton. *Optica*. Translation by Carlos Solís. Alfaguara. Madrid, 1977.
24. A. Koyré. 'The significance of the Newtonian synthesis'. *Archives Internationales d'Histoire des Sciences*, **3**, [29] 1954, pp. 291–311.
25. E. Halley, 'Ode dedicated to Newton'. *Op. cit.* (1).
26. R. S. Westfall. *Never at Rest. A Biography of Isaac Newton*. Cambridge University Press. Cambridge, 1977.
27. I. B. Cohen. 'Newton's discovery of gravity'. *Scientific American*, March, 1981, pp. 123–133.

28. R. S. Westfall. 'The changing world of the Newtonian industry'. *J. Hist. of Ideas*, **37**, 1976, pp. 175–184.
29. B. J. Dobbs. *The Foundations of Newton's Alchemy, or 'The Hunting of the Greene Lyon'*. Cambridge University Press. Cambridge, 1975.
30. B. J. Dobbs. 'Newton's alchemy and his theory of matter'. *Isis.*, **13**, 1982, pp. 511–528.
31. D. Castillejo. *The Expanding Force in Newton's Cosmos*. Ed. de Arte y Bibliofilia. Madrid, 1981.

CARLOS TORRES

THE PHILOSOPHY AND THE PROGRAMME OF HILBERT

INTRODUCTION

This essay is due to my desire to bring the reader interested in mathematical philosophy and foundational studies into closer touch with Hilbert's thought on the subject. Still it is not intended as an introduction to the matters under discussion and some familiarity with them is presupposed (at least some acquaintance with mathematical logic). Rather, it is a synthetic view of what I think is the nucleus of Hilbert's mathematical philosophy and its unfolded consequences.

Those ideas considered rudimentary in Hilbert's conception of mathematics are presented by a 'thesis' outline which is introduced simply for economy. I do not claim that these theses were Hilbert's own. Rather, I derived them from his writings and those of his commentators, but they are also to be found in his papers. They summarize his opinion with regard to the essence of mathematical thought as it is spread throughout his writings, and represent the core of his conception on the matter. Perhaps they are misleading, but nevertheless they reflect my personal comprehension of his view. The following text should be read bearing in mind these few principles.

The following text is divided into a series of fragmented remarks where the theses are unfolded, commented upon or simply explained. Hilbert's programme is expressed as a concrete means to attain the goal of founding mathematics on perception.

Inasmuch as Gödel's work plays a fundamental role in the mathematical response to the programme, two full sections deal with his theorems. A modern sketch of these is set forth on the basis of the Diagonal Lemma and the Hilbert and Bernays derivability conditions. After a brief survey of the effect of Gödel's work, we come to a hasty end. After all, not much is left to be said on the mathematical perspective: there is no way to achieve the objectives posed by the programme; it points toward an unattainable solution. This response to the programme is mathemati-

S. Ramirez and R. S. Cohen (eds.), Mexican Studies in the History and Philosophy of Science, 151–172.

cal, not philosophical. Properly interpreted, it means that a pure 'sign' intuition, as Hilbert proposes, is not enough to found mathematics.

I acknowledge my indebtedness to the works of J. Herbrand, S. Körner, C. Reid, C. Smorynski, G. Boolos & R. Jeffrey, H. Weyl, J. Ladiere, J. von Neumann, and G. Kreisel. I have borrowed several ideas from them, sometimes without special mention.

SYNOPSIS

Hilbert posed the following philosophical project: to found mathematics upon pure sign intuition. In order to achieve it, he expounded a mathematical programme.

(1) THE NATURE OF MATHEMATICS

Thesis #1. Mathematics comprises two kinds of notions: descriptive and ideal. *Grosso modo*, descriptive notions correspond to concrete objects and constructions of sense-experience. On the contrary, ideal notions are ideas of reasons which transcend every experience. Their function is to supplement mathematical reason. They are, in this sense, indispensable.

Thesis #2. Descriptive notions have their origin in certain kind of intuitive reasoning which is at the base of mathematical thought. This reasoning, which proceeds without any axiomatic assumptions, consists of direct reflection upon objects that are present in intuition as immediate experience, previous to any thought. It constitutes what is called 'finitary mathematics'. Soon this mathematics is extended through the addition of ideal notions and sentences that deal with fictitious objects not constructible in perception. Examples are Cantor's transfinite numbers and the notion of the actual infinite.

Thesis #3. The incorporation of ideal notions in a theory that originally deals with concrete objects and constructions is bound only to one condition: that of being consistent (non-contradictory). At this point there is agreement with Kant.

Thesis #4. The only way to give form to a mathematical theory which includes ideal notions and inferences is to proceed axiomatically. In fact it is such a procedure that is specific to mathematical thought.

Thesis #5. The only requirement for an axiomatic theory is that of being non-contradictory (consistent). In that event, its axioms are true and the objects defined by them exist.

1. For Hilbert the occurrence of contradictions within Cantor's transfinite theory does not force him to abandon it. In his opinion, antinomies

are the result of the carelessness with which finitary mathematics has been extended with the introduction of ideal notions. His motto: *No one shall drive us out of the paradise which Cantor has created for us.*

2. For Hilbert, ideal statements cannot be justified as evident material truths. By the same token, transfinite forms of deduction are not justified by the intuition one has of them.[1] The only justification will be the proof that their assumption will not lead to contradiction.

3. Hilbert holds that in the development of an axiomatic theory, the nature of the objects considered is irrelevant. The only important things are the relations fixed by the axioms among said objects. Thus, in his opinion, mathematics is not a way of attaining a supra-sensible reality. In fact, the objects with which a theory deals do not need to exist beyond that theory.

4. In Hilbert's hands, mathematics becomes the science of that which is possible. By 'possible' is meant that which does not lead to contradiction. For Hilbert, an axiomatic theory is significant if one can prove that no contradiction can be deduced from its axioms. Such is the essential problem for the foundation of mathematics as he understands it in his later work.

5. To develop a theory which includes ideal and descriptive notions in a legitimate manner, it is necessary to prove its consistency *a priori*. A fundamental task is thus imposed: to look for the guarantee of the consistency of classical mathematics, including Cantor's transfinite number theory. For such a purpose one needs to:

(a) Axiomatize classical mathematics.
(b) Prove the consistency of such an axiomatic system.

Between 1908 and 1922, Zermelo and Fraenkel developed an axiomatic system for set theory which in general fulfills the first task. Its consistency was still at stake.

6. Hilbert regards the problem of consistency as the central question in the foundation of mathematics. To cope with it, he advances a new method which has at its base the literal sense of the word 'consistency': a contradiction cannot be deduced (i.e. there are no two theorems, one of which is the negation of the other).

(2) THE PHILOSOPHY OF SIGN

Thesis #6. The subject matter of mathematics is given independently of logic. Hence it cannot be founded on logic alone. Rather, the use of logical deduction requires that something is given in representation (*Vorstellung*) as an immediate experience prior to any thought. In accordance with this theory, the subject matter of mathematics are the concrete signs whose forms are clear and immediately recognizable. These objects can be thoroughly examined in all of their parts: their properties, their differences, and their order are intuitively given with themselves as something that cannot be reduced to something else and that does not require such a reduction.

Thesis #7. The firm base for mathematical thought is sign intuition. Upon it, consistency proofs are to be sustained.

Thesis #8. Within the sphere of sense experience no contradiction can have a place. Contra-*diction* can only take place in *dictum*, that is, among statements.[2]

Thesis #9. Finitary mathematics is, by its own nature, non-contradictory because its statements describe objects and the relations among them in a precise manner.

7. Mathematical reality is not placed in an ideal world but rather within the sphere of what is perceived or of what can be given in perception. Its objects, signs, have an existence independent of thought and of the constructions through which one comes to know them. These objects do not mean nothing by themselves.

8. Signs are, as concrete objects, previous to any thought and irreducible to anything else.

9. For Hilbert, the only safe haven for mathematics is sign intuition (which, in Ladrière's words, "enjoys (being) privileged evidence"). In Hilbert's opinion, the validity of mathematical thought is sustained on two facts: upon the objective and independent existence of signs and its perfect adequation to the reality (that of concrete signs) that is its subject matter.

10. The renewal of sign intuition forces the reduction of mathematics to the domain of sensible perception.

(3) THE PROBLEM OF CONSISTENCY. FORMALIZATION

11. From the above, we see that, for Hilbert, the definite solution to the problem of the foundations of classical mathematics is achieved through a finitary proof of its consistency. To carry out this task, axiomatization is not enough: an axiomatic theory is an ill-defined object of thought, and it is an essential part of finitary thought that it can only be applied to entities that are constructively given and defined. The representation of

theory in the realm of perception and the study of its deductive structure are not possible at this level. There is only one way to cope with these requirements, namely formalization.[3]

12. In order to completely formalize classical mathematics, it is necessary (Von Neumann, 1931):

(a) To enumerate all the symbols used in mathematics and logic. These symbols, called 'primitive symbols', include the symbols '$\sim$' and '$\rightarrow$' (which stand for 'negation' and 'implication' respectively).
(b) To characterize unambiguously all the combinations of these symbols which represent statements classified as 'meaningful' in classical mathematics. These combinations are called 'formulas'.
(c) To supply a construction procedure which enables us to construct successively all the formulas which correspond to the *true* statements of classical mathematics. This procedure of formal derivation is called 'proving' and the formulas constructed with it are called 'theorems'.

The latter is achieved by defining 'proving' as follows:

(c_1) Certain formulas, characterized in an unambiguous and finitary way are called 'axioms'. Every axiom is considered proved.
(c_2) Certain rules, characterized in an unambiguous and finitary way are called 'inference rules'. These indicate how to move from a group of formulas called 'hypotheses' to another formula called 'conclusion'. We say that the conclusion is 'inferred' from the hypotheses.
(c_3) Defining 'proof' as a finite sequence of formulas, each of which is an axiom or is inferred from previous ones in the sequence through the application of any of the inference rules. The last formula in a proof is called 'theorem'.[4]

These tasks had already been carried out by the efforts of the logicists. In *Principia Mathematica* (1910–1913) Russell and Whitehead provide massive experimental evidence that classical mathematics can be represented in a formal system[5] with a very small number of symbols and inference rules. Even if *Principia* was created in a very different context, Hilbert adapts it for his own purposes by denying any possible meaning to its symbols, declaring that the formulas of the logical calculus in *Principia Mathematica* are ideal statements which mean nothing in themselves.[6]

Thus, in lieu of a mathematical science expressed in ordinary language, one gets an inventory of formulas, a system of signs that combine and dissociate according to objective rules which are precisely described. Some of these correspond to mathematical axioms. Rules correspond to the material inferences according to which formulas are linked.

13. Through formalization, Hilbert is capable of reducing abstract mathematical proofs to the realm of perception. In it he places his hopes for a solution to the problem of consistency. He considers that even if it is true that every formal theory that represents classical mathematics must represent the ideal notions of which it makes essential use, it is also true that the constructive character of formal derivations allows one to attack the problem from a finitary viewpoint; in other words, even if the intuitive interpretation of a formal system is not finitary, the system, seen as a totality of derivations (i.e. finite sequences of ordered formulas according to the rules), is constructively defined and can be the matter of finitary argumentation. In this manner, in Hilbert's opinion nothing will prevent validation of the mathematical apparatus.

14. The formalization of classical mathematics must be *complete.* That is, the theorems of both theories (formal and axiomatic) must correspond to each other in an obvious sense. Otherwise the formalization will be inadequate.

(4) METAMATHEMATICS

15. Formalization confronts mathematical thought with a system of objects, signs, in which axioms, formulas, and formal derivations are the subject of an intuitive examination. This is carried out within a discipline that could, in a sense, be considered new: *metamathematics.* This includes the intuitive considerations that are brought forth by formal systems. Its purpose is to establish the non-contradictory character of the axioms.

16. It is within metamathematics that the description of formal systems takes place. Here the purpose of statements is to *communicate.* Its rudiment is sign intuition.[7] The *leit-motif* of this discipline is the syntactical analysis of formal systems in order to characterize, in objective terms, the aggregate of its theorems. Hilbert, in particular, tries to prove that certain formulas are not theorems (consistency) and, if possible, to

find a procedure that will allow us to decide whether a given formula is – or is not – a theorem (a decision method).

17. Metamathematics, as understood by Hilbert, is essentially syntactic: the truth of its statements depends exclusively on the form of the expressions under consideration and not on their meaning.

18. Only finitary arguments and modes of reasoning are admitted in metamathematics. The reason is obvious: if what is to be justified are the transfinite elements of classical mathematics, it does not make sense to appeal to them. The tools must be kept pure. Some of the unacceptable modes of reasoning are: existential proofs that appeal to *reductio ad absurdum*, transfinite induction, the use of the axiom of choice and its equivalents, any use of an actual infinite, any operation that makes an infinite number of manipulations of formulas, or an appeal to an infinite number of structural properties of the formal system. Essentially, the point is to forbid any deductive step that is not justified by intuitive evidence.

19. For Hilbert, *to comprehend* the concept of infinity is *to understand* the use that is made of it.[8] Such a use is, in his opinion, the dynamic of the development of the theory. Its use is justified when proving that any finitary statement that is obtained with its aid is true. This would happen, according to Hilbert, if classical mathematics was consistent, because the deduction of a false finitary statement S would lead directly to a contradiction as its negation $\sim S$ can be verified with finitary methods.[9]

20. Hilbert wants to solve the problem of consistency through formalization. To the latter he assigns the role of an instrument in the investigation of the deductive structure of preexisting mathematical theories. In this sense it is erroneous to think that he is only making reference to signs, that he would be a defender of strict formalism. Formalization is but a means to eliminate the transfinite machinery, a representation.

21. There is no precise definition of the concepts of 'finitary proof' and 'formalization'. Nevertheless, finitary reduction needs no clarification. Once the job is done, an examination of the methods utilized will reveal whether they are satisfactory or not.

(5) THE PROGRAMME

22. Hilbert's programme can be summarized as follows:

(1) To formalize classical mathematics.
(2) To prove that the formalization is (semantically) complete.[10]

(3) To prove with finitary methods that the formal system is consistent.

23. Hilbert reduces the problem of consistency to proving that there is a formula that is not a theorem in the system.[11] This last problem is the subject of an intuitive treatment: what is pursued is to prove the impossibility of producing certain kinds of derivations. The situation is similar to that which comes to pass in elementary number theory when one proves that $\sqrt{2}$ is irrational. In this case the problem is to prove that it is impossible to find two positive integers a and b which stand in the relation $2a^2 = b^2$ (i.e. it is also a case of an impossibility, to produce two numerical symbols with a certain property). A normal proof, as well as a numerical symbol, is a visible concrete object. The condition that the last formula of the proof must not be a predetermined formula is something that can be verified by direct inspection. As Cavailles puts it, "a written reasoning cannot deceive us, since the excluded figures would appear in its design" (Cavailles, 1937).

24. The *mathematical* character of Hilbert's programme must be stressed.[12] It is a *mathematical* examination of logic in its application to mathematics with the aim of justifying its methods.

This shows the *mathematical* character of his philosophy. In this sense we can follow Cavailles in saying that Hilbert's philosophical position propounds a philosophy *derived from mathematics, a mathematical philosophy*, rather than *a philosophy of mathematics* that would endeavor to adjust the concrete work of working mathematicians to positions decided beyond this practice.[13]

The programme is, at the same time, a mathematical expression of a philosophical project which seeks to found mathematics upon what is perceived or can be given in perception.

25. The Programme was, *prima facie*, reasonable since the proofs of consistency seemed to depend only on the deductive structure of the system and its combinatorial character.

(6) GÖDEL

26. The aim of *proof theory*[14] was to "eliminate forever the problem of foundations". As a preliminary step to the proof of consistency of classical mathematics, Hilbert and his followers put their tools to the test by applying them to logic itself and to certain fragments of arithmetic. The first step was to formalize first order logic and to prove its consistency

with finitary methods. After this, they directed their attention to number theory. In this case, formalization was staged by considering successive fragments of Peano's arithmetic. In each case, a finitary proof of consistency was provided. By 1930, there was enough reason to expect that soon success would crown the effort. Let us now listen to H. Weyl:

> But such bright hopes were dashed by a discovery in 1931 due to Kurt Gödel, which questioned the whole program. Since then the prevailing attitude has been one of resignation. The ultimate foundations and the ultimate meaning of mathematics remain an open problem; we do not know in what direction it will find its solution, nor even whether a final objective answer can be expected at all. 'Mathematizing' may well be a creative activity of man, like music, the products of which not only in form but also in substance are conditioned by the decisions of history and therefore defy completely objective rationalization. (...)
>
> Gödel showed that in Hilbert's formalism, in fact in any formal system M that is not too narrow, two strange things happen: (1) One can point out arithmetic propositions Φ of comparatively elementary nature that are evidently true yet cannot be deduced within the formalism. (2) The formula Ω that expresses the consistency of M is itself not deducible within M. More precisely, a deduction of Φ or Ω within the formalism M would lead straight to a contradiction in M, i.e. to a deduction in M of the formula $\sim (1 = 1)$. From the first fact one learns that the fields of propositions accessible to insight on the one hand and to deduction on the other overlap, neither of the two being contained in the other. Symbolic mathematics, while in some directions going far beyond what is capable of verification based on evidence, in other directions accomplishes less. Although the idea of a transcendental world existing and complete in itself is the guiding principle in building up our formalism, that formalism at any fixed stage has the character of incompleteness, inasmuch as there will always be problems, even problems of a simple arithmetical nature, that can be formulated within the formalism and decided by insight, but not decided by deduction within the formalism. (...)

> Gödel's second theorem is even more disquieting, for it confronts us with this alternative: either the reasoning by which consistency of the formalism is established must contain some argument that has no formal counterpart within the system (...) or the idea of a strictly 'finitistic' proof of consistency must be given up altogether (H. Weyl, 1949, pp. 219–220).

Gödel shows that metamathematics, as mathematics itself, can be dealt with formally.[15] The virtue of this consists in observing that, from a general point of view, the nature of the objects considered in the construction of a formal system (symbols, formulae, proofs, etc.) is irrelevant. Consequently, these can be replaced by natural numbers. The code for substitution can be made in such a way that the rules of the system, those that govern the formation of formulae and the derivation of theorems, are expressed through simple arithmetical operations among the numbers representing them. Thus, the metamathematical statements become arithmetical and, in turn, are formalizable.

This procedure is known as *arithmetization of syntax*. When applied to a formalization of arithmetic, a peculiar situation is generated: metamathematics acquires the same status as the formalized theory, and its statements can be expressed within the system. In this way, the formalism will contain a representation of its metatheory.[16]

(7) THE THEOREMS

28. We cannot skip a more detailed explanation of Gödel's theorems. Its proof for a formal system M rests on the possibility of encoding (i.e. representing through a code) part of the metatheory about M in M itself. This reveals the following cycle:

(1) Through arithmetization, certain metatheoretical properties of M (originally formulated in English) are expressed in arithmetical form.
(2) Through encoding, the arithmetical statements which express properties about M are formalized in M.
(3) Through arithmetization, certain formulae in M, originally interpreted as arithmetical statements, are now interpreted as metatheoretical statements.

Thus, it is now possible to find formulae that (through arithmetization) are interpreted as referring to formal expressions that coincide with themselves. In this manner the possibility of finding self-referent formulae is established (through arithmetization). In the proof of the Diagonal Lemma we shall see how such formulae are constructed, and in the subsequent theorems (i.e. Gödel's theorems), what can be done with them.

29. A detailed proof of these theorems is not given here. Thus a description of M is not given, nor an explanation about the manner in which the encoding of its metatheory is carried out. Rather we assume that the formal system M satisfies conditions that will be listed below.[17]

Condition one. There is a one-to-one function $g\colon C \to \mathrm{N}$ with the following properties:

(1) For every $c \in C$, the value $g(c)$ is computable through a finitary procedure.
(2) There is a finitary procedure to decide whether a natural number n is the image of a constituent of M. In case there is a constituent c whose image is n, the procedure allows its effective determination.

g is called a Gödel correspondence. The number $g(c)$ associated to each constituent $c \in M$ is the Gödel number of c. Through a Gödel correspondence, metamathematical propositions about the constituents of M become arithmetical propositions about its Gödel numbers. This is the basis for the *arithmetization* of the syntax of M.

Condition two. M is provided with *numerals* (closed terms) $\bar{0}$, $\bar{1}$, $\bar{2}$ etc. which represent the natural numbers 0, 1, 2, etc.[18]

If c is a constituent of M with Gödel number g, then $\ulcorner c \urcorner$ denotes the numeral $\bar{g}$ (i.e. $\ulcorner c \urcorner = \overline{g(c)}$). In what follows, the expression $\ulcorner c \urcorner$ serves as a code (a name) for c in M. In particular, to each formula A of M there corresponds a closed term $\ulcorner A \urcorner$ which is called the *code* of A.

Condition three. M is consistent.[19]

Condition four. There is a term $sb(x, y)$ in M with the following property: if $A(x)$ is a formula whose only free variable is x and t is a term of M, then $\vdash sb(\ulcorner A(x) \urcorner, \ulcorner t \urcorner) = \ulcorner A(t) \urcorner$ (i.e. it is proved in M that $sb(\ulcorner A(x) \urcorner, \ulcorner t \urcorner)$ is the code of the formula obtained when substituting, in $A(x)$, t for the free occurrence of x).

Condition five. There is a binary predicate $PR(x, y)$ in M which satisfies the following:

(1) If h is the Gödel number of a proof of the formula with Gödel number k, then $\vdash PR(\bar{h}, \bar{k})$.

(2) If h is not the Gödel number of a proof of the formula with Gödel number k, then $\vdash \sim PR(\bar{h}, \bar{k})$.[20]

Condition six. The predicate $Th(y) \equiv \exists x PR(x, y)$ satisfies the following *derivability conditions* (Hilbert and Bernays, 1934):

(D_1) If $\vdash A$, then $\vdash Th(\ulcorner A \urcorner)$

(D_2) $\vdash Th(\ulcorner A \urcorner) \rightarrow Th(\ulcorner Th(\ulcorner A \urcorner) \urcorner)$

(D_3) $\vdash Th(\ulcorner A \urcorner) \wedge Th(\ulcorner A \rightarrow B \urcorner) \rightarrow Th(\ulcorner B \urcorner)$,

where A and B are statements of M.

The formula '$Th(y)$' represents in M the syntactic property 'y *is a theorem of* M'. Nevertheless, it is not always the case that $\vdash A \Leftrightarrow \vdash Th(\ulcorner A \urcorner)$. The reason is that the existential quantifier that appears in the formula transforms it into an *ideal statement*, and it can happen that the system, even being consistent, might prove false ideal statements (i.e., that it happens that $\vdash Th(\ulcorner A \urcorner)$ and $\nvdash A$ for some statement A).

Lemma 1. If M is ω-consistent and $\vdash Th(\ulcorner A \urcorner)$, then $\vdash A$.

Proof

Suppose that $\nvdash A$. In this case the statements

'0 *is not a proof of the formula* $g(A)$'

'1 *is not a proof of the formula* $g(A)$'

'2 *is not a proof of the formula* $g(A)$'

(one for each $k \in \mathbb{N}$) are all true. As $PR(x, y)$ codifies proofs of M in M, one has that

$$\vdash \sim PR(\bar{0}, \ulcorner A \urcorner)$$
$$\vdash \sim PR(\bar{1}, \ulcorner A \urcorner)$$
$$\vdash \sim PR(\bar{2}, \ulcorner A \urcorner)$$

(a formula $\sim PR(\bar{k}, \ulcorner A \urcorner)$ for each $k \in \mathbb{N}$) and since the system is ω-consistent, then $\nvdash \exists x PR(x, \ulcorner A \urcorner)$ (i.e. $\nvdash Th(\ulcorner A \urcorner)$). Therefore, if

$\vdash Th(\ulcorner A \urcorner)$ then $\vdash A$.

Lemma 2 (Diagonal Lemma). Let $A(x)$ be a formula whose only free variable is x. Then there is a *sentence* (formula without free variables) B such that $\vdash A(\ulcorner B \urcorner) \leftrightarrow B$ (i.e., there is a *fix point* for $A(x)$).

Proof

Let $D(x) \equiv A(sb(x,x))$ the diagonalization of $A(x)$, $d = \ulcorner D(x) \urcorner$ and $B \equiv D(d)$.

$\vdash B \leftrightarrow D(d)$	by definition of B
$\vdash D(d) \leftrightarrow A(sb(b,d))$	by definition of D
$\vdash A(sb(d,d)) \leftrightarrow A(sb(\ulcorner D(x) \urcorner, d))$	by definition of d
$\vdash A(sb(\ulcorner D(x) \urcorner, d)) \leftrightarrow A(\ulcorner B \urcorner)$	$\vdash sb(\ulcorner D(x) \urcorner, d) = \ulcorner D(d) \urcorner$ by condition four.

and

$\vdash B \leftrightarrow A(\ulcorner B \urcorner)$.[21]

First incompleteness theorem (Gödel, 1931). If $\vdash G \Leftrightarrow {\sim} Th(\ulcorner G \urcorner)$,[22] then:

(1) $\nvdash G$
(2) $\nvdash {\sim} G$ (assuming the ω-consistency hypothesis).

Proof of (1)

1. G	theorem by hypothesis
2. $G \rightarrow {\sim} Th(\ulcorner G \urcorner)$	theorem by definition of G
3. ${\sim} Th(\ulcorner G \urcorner)$	immediate consequence of 1 and 2
4. $Th(\ulcorner G \urcorner)$	from 1 by D_1
5. ${\sim} Th(\ulcorner G \urcorner) \wedge Th(\ulcorner G \urcorner)$	logical consequence of 3 and 4

and M would be inconsistent.

Proof of (2)

1. ${\sim} G$	theorem by hypothesis
2. ${\sim} Th(\ulcorner G \urcorner) \rightarrow G$	theorem by definition of G
3. ${\sim} G \rightarrow Th(\ulcorner G \urcorner)$	contrapositive of 2

4. $Th(\ulcorner G \urcorner)$ — immediate consequence of 1 and 3
5. G — from 4 by Lemma 1
6. $G \wedge \sim G$ — logical consequence of 1 and 4

and M would be ω-inconsistent.

Let CON be the formula $\sim Th(\ulcorner G \wedge \sim G \urcorner)$. Metamathematically the formula CON asserts the consistency of M when stating "*the contradiction $G \wedge \sim G$ is not a theorem of M*".

Lemma 3. $\vdash G \leftrightarrow CON$.

Proof of $\vdash G \rightarrow CON$

1. $G \rightarrow \sim Th(\ulcorner G \urcorner)$ — theorem by definition of G
2. $G \vee \sim G$ — theorem: law of excluded middle
3. $(G \vee \sim G) \Rightarrow (G \wedge \sim G \Rightarrow G)$ — theorem: tautology $A \rightarrow (\sim A \rightarrow B)$
4. $G \wedge \sim G \rightarrow G$ — immediate consequence of 2 and 3
5. $Th(\ulcorner G \wedge \sim G \rightarrow G \urcorner)$ — from 4 by D_1
6. $Th(\ulcorner G \wedge \sim G \urcorner) \rightarrow Th(\ulcorner G \urcorner)$ — from 5 by D_3
7. $\sim Th(\ulcorner G \urcorner) \rightarrow \sim Th(\ulcorner G \wedge \sim G \urcorner)$ — contrapositive of 6
8. $G \rightarrow CON$ — from 7: $G \leftrightarrow \sim Th(\ulcorner G \urcorner)$ and $CON \leftrightarrow \sim Th(\ulcorner G \wedge \sim \urcorner)$

Proof of $\vdash CON \rightarrow G$

1. $Th(\ulcorner G \urcorner) \rightarrow Th(\ulcorner Teo(\ulcorner G \urcorner) \urcorner)$ — theorem by D_2
2. $G \rightarrow \sim Th(\ulcorner G \urcorner)$ — theorem by definition of G
3. $Th(\ulcorner G \urcorner) \rightarrow \sim G$ — contrapositive of 2.
4. $Th(\ulcorner Thh(\ulcorner G \urcorner) \urcorner) \rightarrow Th(\ulcorner \sim G \urcorner)$ — from 3 by D_1 and D_3
5. $Th(\ulcorner G \urcorner) \rightarrow Th(\ulcorner \sim G \urcorner)$ — transitivity of $\rightarrow$ to 1 and 4
6. $\sim G \rightarrow (G \rightarrow G \wedge \sim G)$ — theorem: tautology $A \rightarrow (\sim A \rightarrow B)$
7. $Th((\ulcorner \sim G \rightarrow (G \rightarrow G \wedge \sim G) \urcorner))$ — from 6 by D_1
8. $Th(\ulcorner \sim G \urcorner) \rightarrow Th(\ulcorner G \rightarrow G \wedge \sim G \urcorner)$ — from 7 by D_3

9. $Th((\ulcorner G \rightarrow \wedge \sim G\urcorner) \rightarrow (Th(\ulcorner G\urcorner) \rightarrow Th(\ulcorner G \wedge \sim G\urcorner))$	D_3
10. $Th(\ulcorner \sim G\urcorner) \rightarrow (Th(\ulcorner G\urcorner) \rightarrow Th(\ulcorner G \wedge \sim G\urcorner))$	transitivity of $\rightarrow$ to 8 and 9
11. $Th(\ulcorner G\urcorner) \rightarrow (Th(\ulcorner G\urcorner) \rightarrow Th(\ulcorner G \wedge \sim G\urcorner))$	transitivity of $\rightarrow$ to 5 and 10
12. $Th(\ulcorner G\urcorner) \rightarrow Th(\ulcorner G \wedge \sim G\urcorner)$	logical consequence of 11
13. $Th(\ulcorner G \wedge \sim G\urcorner) \rightarrow Th(\ulcorner G\urcorner)$	contrapositive of 12
14. $CON \rightarrow G$	$G \leftrightarrow \sim(\ulcorner G\urcorner)$ and $CON \leftrightarrow Th(\ulcorner G \wedge \sim G\urcorner)$.

Second incompleteness theorem (Gödel, 1931). $\nvdash CON$.

Lemma 3 shows that Gödel's statement is equivalent to the formula CON that 'asserts' the consistency of M. The underivability of CON follows from the underivability of G, as was proved in the first theorem. By the way, CON can be any formula of the form $\sim Th(\ulcorner \sim H\urcorner)$, where H is a theorem of M.

30. Gödel's first theorem was modified in 1936 by Barkley Rosser in such a way that the existence of undecidable statements follows from the sole hypothesis of consistency. Thus it is shown that the stronger hypothesis of ω-consistency is not necessary.

31. Gödel–Rosser's theorem shows that no formal system M for arithmetic can be consistent and complete at the same time. In any of them a statement G can be produced that is (arithmetically) true but that cannot be proved or disproved in M in case M is consistent. The proof is effective in the sense that it is known how to produce the undecidable statement G.

32. Incompleteness is a fault inasmuch as the system is incapable of proving all the true statements that can be framed within itself. This limitation is inherent to every formal system with a certain expressive power and cannot be avoided.[23]

(8) CONSEQUENCES FOR THE PROGRAMME

33. The first task to be achieved by the programme was to formalize classical mathematics in a formal system in which to every true mathematical statement there would correspond a theorem. Syntactic com-

pleteness of the system would correspond to this property. As Gödel and Rosser show, such a task cannot be carried out. Hilbert's belief in the existence of a system of such a nature had no basis.[24]

34. Gödel's second theorem shows that the consistency of formal arithmetic can only be proved with methods that cannot be formalized in the system. This difficulty is an almost insurmountable obstacle for the programme, since the finitary methods proposed by Hilbert are essentially arithmetical: they have as their principle the intuitive consideration of elementary combinations of signs and, for this reason, they can be expressed in arithmetic. This makes the possibility of finding a finitary proof of the consistency of classical mathematics very unlikely: it is not clear how a finitary proof would – as the one Hilbert wants – not be formalizable in a system for classical mathematics.

35. The possibility of a finitary proof is not excluded by Gödel's second theorem. In 1931 he wrote:

> I wish to note expressly that theorem XI (and the corresponding results for *M* and *A*) do not contradict Hilbert's formalist viewpoint. For this viewpoint presupposes only the existence of a consistency proof in which nothing but finitary means of proof is used, and it is conceived that there exist finitary proofs that *cannot* be expressed in the formalism of *P* (or *M* or *A*) (Gödel, 1931).

Even if the theorem is not conclusive with respect to the possibility of a finitary proof of consistency, it must be said that nothing has been done in this field. At any rate, this argument stems from the imprecision of the notion of 'finitary proof'. It could have a positive result only if a proof of consistency were shown as evidently finitary.

36. Gödel tries to offer other approaches to Hilbert, at least with respect to consistency proofs. In a paper called 'Über eine bisher noch nicht benutzte Erweiterung des finiten Standpunktes' he points towards the need to widen Hilbert's finitary point of view:

> P. Bernays has repeatedly pointed out that, in view of the fact that we cannot prove the consistency of a system by methods of proof which are restricted enough to be represented within the system itself, we have to step outside of finitary mathematics (in Hilbert's sense) in order to prove the consistency of classical mathematics, or even the consistency of classical number theory. Since finitary mathematics is de-

> fined as that which rests on inspectable evidence, this means ... that in order to prove the consistency of number theory we need certain abstract concepts. For this purpose we must count as abstract (not inspectable) those concepts which are essentially of the second or higher order. By this we mean those concepts which do not comprise properties or relations of concrete objects (as for instance combinations of symbols), but which are concerned with thought-constructions (proofs, meaningful propositions, etc.). The consistency proofs will use insights into these thought-constructions, insights which are derived not from the combinatorial (space–time) properties of the combinations of symbols which represent the proofs, but only from the sense of the symbols (Gödel, 1958).

Upon a widening of this sort, Gödel offers a proof for the consistency of intuitionist arithmetic and, hence, of classical arithmetic (which he proved equivalent to the former in 1932). Of what value can consent to such proofs of consistency be? Evidently, an inferior value to what would be given to a finitary proof. In particular, Gödel's proof is not finitary as it deals with non-concrete objects (the so-called *computable functions of finite type*). It is a subjective matter to accept or reject his methods as satisfactory. Such is the cost to be paid for consistency proofs: lesser intuitive evidence. As H. Weyl points out, "The boundary line of what is intuitively trustworthy has once more become vague" (H. Weyl, 1944).

37. In what concerns the consistency of classical mathematics, there is no hope that it can be shown with known methods. Notwithstanding, Hilbert cherished until his death the illusion of achieving it. He held that even if it were necessary to make recourse to concepts external to the system, these would be finitary, intuitively concrete and acceptable. His attitude shows the unbounded optimism that always characterized him, his unlimited trust in the power of human understanding.

EPILOGUE

Apropos of Hilbert's programme Kreisel wrote:

> As far as an over-all philosophical understanding is concerned, the original Hilbert's programme has failed and, as is usual with great schemes, it gives no hint of what might take its

> place. When asked: "what is mathematics about?", Hilbert could still have said: about the arithmetico-combinatorial facts of finitist mathematics. Even though the latter may raise problems of their own, such a "reduction" could have been satisfying. Hilbert's answer is simply not true even for the very weak sense of "equivalence of content" expressed in statements of formal deducibility and nondeducibility. Furthermore, we have no idea in what sort of investigation we could even hope to find a satisfactory answer to such a question (G. Kreisel, 1958).

Hilbert's answer is false inasmuch as mathematical reasoning cannot be reduced to the finitary. As Gödel shows, it is neither possible to represent the whole aggregate of existing mathematics in the frame of an appropriate formalism, nor to give an account of certain properties of formal systems when the methods of proof are restrained to finitary ones. This raises the question of whether Hilbert's programme can be partially fulfilled, i.e. whether a *portion* of infinitary mathematics can be justified by means of a finitary proof of consistency. It also points out that it is necessary to use methods other than finitary for a significant approach to foundations.[25] Taking for granted that a finitary reduction is not possible, the question arises: to which constructive theories can classical mathematics be reduced? The pursuit of such a *reductive programme* is the core of actual research on foundations (W. Siege, 1988). However, contrary to Hilbert's aim, it does not have the solution of the problem of foundations once and for all as its purpose. Rather it is foundational research where the relation between constructive and non-constructive mathematics is explored.

In regard to philosophy we notice at once that recourse to pure sign intuition was not enough. Hilbert's expectation, namely *to found mathematics upon that which can be given in perception*, was a delusion. It remained an open problem whether there is an epistemological principle upon which mathematics can be grounded.

NOTES

[1] A case of an ideal notion is the law of the excluded middle when it is applied to infinite sets. Given a property $P(x)$ and an infinite set A, is it legitimate to assert that there is an x such that $P(x) V \sim P(x)$? Any positive answer to this question lacks an empirical base. Nevertheless, most mathematicians accept this law as valid for all sets,

not only for finite ones. This ideal notion is introduced for the purpose of reestablishing the validity of the laws of Aristotelian logic that were lost, as Brouwer shows, during the transition to infinity.

[2] In Spanish, the verb to say is *decir* and the word for contradict is *contradecir*. The last part of the sentence is based on a language game which has no literal translation: '*Solo los enunciados se contra-DICEN*'.

[3] Formalization is an involuntary step produced by the requirements of the problem of foundations. It is introduced in order to provide the deductive methods of mathematics a form so exact as to become objects of *mathematical* research (consistency is a feature relative to the proofs of a theory and not to its statements). This difficulty is not necessarily a hindrance. It is also the completion of intense work in axiomatics.

[4] There is a strong resemblance between an axiomatic theory and its formalizations. In an axiomatic theory *statements* are enunciated within a natural language and are about technical terms which remain undefined. Among said statements some are held to be *true* and some others as *false*. The true statements of the theory are the axioms and the theorems. An *axiom* is a statement deliberately chosen as unproved, and a *theorem* is a statement which is inferred from the axioms through *logical deduction* (while a *false* statement is one whose negation is true). All these elements have an intuitive sense. In a formalism there are analogous elements, but their meaning is rigorously determined by the rules of the system. The notion of *formula* corresponds to the notion of statement in the axiomatic theory, the notion of *proof* to that of logical deduction, the notion of *theorem* to that of true statement, and the notion of *refutable* formula to that of false statements.

[5] 'Formal theory', 'formal system', 'logical calculus' and 'formalism' are but different names for the same concept. The same can be claimed for the words 'sign' and 'symbol', even if the latter can mean 'sequence of signs'. On the other hand, the word 'expression' means 'finite sequence of primitive symbols'.

[6] Within a formalism, signs are no longer the representation of something, not a picture that would elicit in understanding the symbolized thing. Through a radical change of attitude, thought is forced to stop in them and to take them for its final objects.

[7] This intuition, among other things, allows recognition of the system signs and to decide whether a given one is one of them. It also allows us to recognize those language expressions that are formulas and among these those which are axioms. By the same token, it permits us to realize whether a finite string of formulas is a proof and it allows us to check if it is correct.

[8] "Operating with the infinite can be made certain only by the finitary" (Hilbert, 1925). In Hilbert's opinion, ideal notions (or in Kantian jargon 'the concepts of reason') can only be understood by reasoning about them in a well defined context. In this case, the context is finitary mathematics.

[9] This argument takes for granted that all finitary methods of proof are available in classical mathematics.

[10] Semantically complete: to each true statement in classical mathematics there corresponds a derivable formula in the system. This condition is often replaced by the following: to prove that formalism is *syntactically* complete, that is, to prove that for every formula S without free variables, S or $\sim S$ is a theorem in the system. This definition has the advantage of being entirely syntactic, (it only makes reference to the

formal system and not to its interpretations) and is, thus, preferable. Both conditions are equivalent if the system is *sound*: if A is derivable in the system, then the statement corresponding to it in classical mathematics is true.

[11] Formalism considered by Hilbert is inconsistent if and only if all of its formulas are theorems. The reason is that every tautology of the form $\sim A \rightarrow (A \rightarrow B)$ is a theorem and the rule $A, A \rightarrow B/B$ (*modus ponens*) is an inference rule.

On the other hand, if natural numbers are represented within the system, the 'underivable' formula can be specified in advance. Hilbert chooses a particularly simple one: $\sim (1 = 1)$. Hence, consistency follows from the proof that '$\sim (1 = 1)$' is not a theorem. This proof would be, at the same time, a justification of the method of ideal elements.

[12] The Programme has a well defined sense within mathematics. This is so in spite of the lack of precision of some of the notions that play in it, such as 'finitary proof' or 'formalization'.

[13] Cf. Ramírez, S., 'Three metaphysical theses on mathematical philosophy', elsewhere in this volume.

[14] In a strict sense, proof theory is more general than metamathematics since the latter voluntarily confines itself to finitary methods. Nevertheless, for Hilbert it is but the same theory with two names. We follow suit.

[15] The stage of formalism which lasts until 1930, is rather naive though very creative. Even if it is true that at this time the fundamental notions and techniques have been forged, it is also true that it is somewhat acritical and ignored its limitations. Such a stage comes to an end in 1931 with the publication of Gödel's theorems, where Gödel leads metamathematics to a second stage, more critical and more reflective.

[16] This term is not used by Hilbert. In fact, the only one he uses with the prefix 'meta' is 'metamathematics'. Nevertheless, the sense of the word 'metatheory' is analogous to the above and means 'mathematical theory about a formal system or language'.

[17] The following definition is needed: the primitive symbols, terms, formulae and proofs of M are called *constituents* of M. The set of constituents of M is denoted by 'C'.

It is assumed that the formal system M includes the axioms and inference rules of the predicate calculus. A reader familiar with it will have no difficulty in dealing with this part of the text.

[18] M generally contains a formalization of Peano's arithmetic in a first order language, a constant '$\bar{0}$' which represents the number 0 and a functional symbol 's' that represents the successor function. In such a case, numerals are the expressions '$\bar{0}$', '$s\bar{0}$', '$ss\bar{0}$' etc. that are formed by repeating the symbol 's' before '$\bar{0}$' a finite number of times.

[19] Some definitions:

M is *consistent* $\Leftrightarrow$ there is a formula A that is not a theorem of M.

M is *consistent* $\Leftrightarrow$ no contradiction $A \wedge \sim A$ is a theorem of M.

M is complete $\Leftrightarrow$ for every statement S of M, $\vdash S$ or $\vdash \sim S$.

A formula A is undecidable in M $\Leftrightarrow$ neither A nor $\sim A$ are theorems of M.

M is ω-inconsistent $\Leftrightarrow$ for some predicate $P(x)$, $\vdash \sim P(\bar{0})$, $\vdash \sim P(\bar{1})$, $\vdash \sim (\bar{2})$, etc. (i.e., $\vdash \sim P(\bar{k})$ for each $k \in \mathbb{N}$) and $\vdash \exists x P(x)$.

M is ω-consistent $\Leftrightarrow$ it is not ω-inconsistent.

(Note. The expression '$\vdash \varphi$' is read 'φ is a theorem of M').

[20] It is said that the predicate $PR(x, y)$ codifies proofs of M in M.

[21] In a sense statement B is self-referent, since it turns out to be 'equivalent' to a formula that 'speaks' of it. Nevertheless, it must be pointed out that such self-reference is possible only through arithmetic when interpreting the numeral $\ulcorner B \urcorner$ as a code for B.
[22] The existence of the statement G is guaranteed by the Diagonal Lemma. G is called '*Gödel's statement*'. It is equivalent (in M) to a formula that interpreted in the metatheory of M states "*G is not a theorem of M*".
[23] We cannot dismiss the following commentary by H. Weyl which is not free of dramatic overtones: "We are not surprised that a concrete chunk of nature, taken in its isolated phenomenal existence, challenges our analysis by its inexhaustibility and incompleteness. (...) But it is surprising that a construct created by mind itself, the sequence of integers, the simplest and most diaphanous thing for the constructive mind, assumes a similar aspect of obscurity and deficiency when viewed from an axiomatic angle" (H. Weyl, 1949, p. 220).
[24] Hilbert also believed that every arithmetical truth had an arithmetic proof. This is also shown false by Gödel's first theorem if by 'arithmetical' is meant 'expressible in classical arithmetic'.
[25] G. Kreisel suggests that instead of a single kind of elementary reasoning whereby we understand the use of transfinite symbols, there will be a *hierarchy* of progressively less elementary but constructive methods, i.e. a hierarchy of Hilbert programmes.

REFERENCES

Barwise, J. (ed.) (1977). *Handbook of Mathematical Logic*, Springer.
Benacerref, P. and Putnam, H. (eds.) (1964). *Philosophy of Mathematics: Selected Readings*, Prentice-Hall.
Bernays, P. and Hilbert, D. (1934). *Grundlagen der Mathematik*, Vol. I, Springer.
Bernays, P. and Hilbert, D. (1939). *Grundlagen der Mathematik*, Vol. II, Springer.
Boolos, G. and Jeffrey, R. (1974). *Computability and Logic*, Cambridge University Press.
Cavailles, G. (1937). *Méthode Axiomatique et Formalisme*, Hermann, Paris.
Gödel, K. (1931). 'On formally undecidable propositions of *Principia Mathematica* and related systems, I', translated by J. van Heijenoort in (1967).
Gödel, K. (1958). 'On hitherto unexploited extension of the finitary standpoint', translated by W. Hodges and B. Watson in *The Journal of Philosophical Logic*, **9** (1980), pp. 133–142.
Heijenoort, J. (ed.) (1967). *From Frege to Gödel*, Harvard University Press.
Hilbert, D. (1904). 'On the foundations of logic and arithmetic', translated by J. van Heijenoort in (1967).
Hilbert, D. (1922). 'Neubegründung der Mathematik', in Hilbert (1935) Vol. 3.
Hilbert, D. (1925). 'On the infinite', translated by J. van Heijenoort in (1967).
Hilbert, D. (1927). 'The foundations of mathematics', translated by J. van Heijenoort in (1967).
Hilbert, D. (1935). *Gesammelte Abhandlungen*, Springer.
Körner, S. (1960). *The Philosophy of Mathematics*, Hutchinson & Co. Ltd.
Kreisel, G. (1958). 'Hilbert's programme', in Benacerref and Putnam (1974).

Ladrière, J. (1957). *Les limitations internes des formalismes*, Louvain.
Neumann, J. (1931). 'The formalist foundation of mathematics', in Benacerrat and Putnam (1964).
Reid, C. (1970). *Hilbert*, Springer.
Smorynski, C. (1977). 'The incompleteness theorems', in Barwise (1977).
Siege, W. (1988), 'Hilbert's program sixty years later'. *J. Symbolic Logic*, **53**, no. 2.
Simpson, S. (1988), 'Partial realizations of Hilbert's program'. *J. Symbolic Logic*, **53**, no. 2.
Feferman, S. (1988). 'Hilbert's program relativized: Proof-theoretical and foundational reductions'. *J. Symbolic Logic*, **53**, no. 2.
Weyl, H. (1944). 'David Hilbert and his mathematical work'. *Bulletin of the American Mathematical Society*, **50**.
Weyl, H. (1949). *Philosophy of Mathematics and Natural Science*, Princeton University Press.

CARLOS ALVAREZ JIMENEZ

SOME LOGICAL REMARKS CONCERNING THE CONTINUUM PROBLEM

Cantor's continuum hypothesis states that the power of the linear continuum, the set of all real numbers, is equal to the power of the second class of transfinite numbers, i.e. the set of all countable transfinite numbers. In terms of the cardinal arithmetic this hypothesis states that $2^{\aleph_0}$ is equal to $\aleph_1$. Even though Cantor himself made a great effort to prove the statement, he never succeeded and it remained as a major problem in set theory at the turn of the century.

Since Hilbert's lecture in Paris in 1900, the continuum problem was considered as a major question to be answered by mathematicians. Up to 1938, two main approaches were handled concerning the power of the linear continuum. On the one hand, the question concerning the possibility of finding a complete answer for this problem within axiomatic set theory. On the other hand, the attempts to find a direct solution by analyzing the nature of the set of real numbers. In this paper we shall analyze some aspects concerning the first approach.

At the International Congress of Mathematicians at Paris in 1900, D. Hilbert presented a list of problems which were to show the way for mathematicians in the new century. The first of these problems (in a list of 23) showed the way for the mathematical research of the infinite: it had to solve Cantor's continuum problem:

> The investigations of Cantor ... suggest a very plausible theorem: every system of infinitely many real numbers is either equivalent to the assemblage of natural integers or to the assemblage of all real numbers and therefore to the continuum. From this theorem it would follow at once that the continuum has the next cardinal number beyond that of the countable assemblage.
>
> ... Let me mention another remarkable statement of Cantor's which stands in the closest connection with the theorem mentioned and which perhaps offers the key to its proof: ... whether the continuum can be considered as well ordered

S. Ramirez and R. S. Cohen (eds.), Mexican Studies in the History and Philosophy of Science, 173–186.

> assemblage – a question which Cantor thinks must be answered in the affirmative.[1]

Two answers were given for this first problem: in 1904, at the Third International Congress of Mathematicians held in Heidelberg, J. König stated that the set of all real numbers could not be well ordered. On the other hand, and after finding an error in König's proof (based on a wrong application of one of Bernstein's laws for the exponentiation of alephs), Ernst Zermelo proved that *any* set could be well ordered, including the linear continuum, by using his choice principle.[2] Since Zermelo's axiomatization of 1908, and besides the objection towards his principle of choice, the possibilities of conceiving the continuum as a well-ordered set and to find the *value* for the cardinal number of the continuum became a problem to be solved within the scope of an axiomatic theory.

SKOLEM'S REMARKS

The possibilities of giving a proof about the *aleph* that corresponds to the power of the continuum based on the axioms for set theory was to meet with some difficulties. In 1922, Thoralf Skolem delivered before the Fifth Congress of Scandinavian Mathematicians the discussion of eight points about the axiomatization of set theory given by Zermelo. Three of these points are closely related to the analysis of the power of the continuum:

1. In the first place, the fact that in any axiomatization, set theoretical notions are necessarily relative.
2. In the second place, the claim that Zermelo's system does not give a complete foundation for set theory.
3. Finally, the fact that several domains could be taken as a model for Zermelo's axioms.

The claim about the relativity of the notions of the theory of sets is based upon the following theorem by L. Löwenheim: "If a proposition of the first order is satisfied in an arbitrary domain, then it is satisfied in a countable domain". From this theorem, Skolem proofs the following generalization:

> Let $U_1, U_2, \ldots$ be a countable sequence of first order propositions. If the hypothesis that all of them are simultaneously satisfied is assumed, then they are simultaneously satisfied in

> a sequence of natural numbers through an adequate determination of the class and of the relation symbols appearing in the proposition.

This generalization can be assumed for Zermelo's axiomatic system. Thus a countable domain B, formed by an infinite sequence of symbols $A_1, A_2, A_3, \ldots$, can be taken, where all the axioms of set theory are satisfied. Skolem makes then a remark towards the fact that B is a countable domain: if Z is the countable set of the sequence of numbers, Z is a set of B; and the same is true for $\cup Z$ (the set of all subsets of Z). But $\cup Z$, according to Cantor, is a non-countable set. This seems a paradox since there is a non-countable set, all of whose elements belong to a countable set. But it is dissolved since the term 'non-countable' is defined *within* B. That is to say, '$\cup Z$ is not countable' means that (i) there is in B a one-to-one function $f : Z \rightarrow \cup Z$, and (ii) there is no one-to-one function (also in B) $g : \cup Z \rightarrow Z$. $\cup Z$ is countable since it belongs to the countable system B, but *within* this model, it is not countable. Skolem concludes that the terms finite, infinite, etc. are all relative to the axiomatization. Thus

> axiomatizing set theory leads to the relativity of set theoretical notions, and this relativity is inseparably bounded up with every thoroughgoing axiomatization (...) Upon an axiomatic base, higher infinites exist only in a relative sense.[3]

The second important remark of Skolem states that Zermelo's axioms are in fact unable to give a complete foundation for set-theory. Since a set could be defined by means of the axioms of set theory which is not an 'existing set' in the domain B where these axioms hold. For this purpose Skolem introduces the notion of 'level' of a set: a set M is of the *first level* if its nth union $\mathfrak{S}^n M$ is the empty set.[4] Sets of the *second level* are those which are not of the first level but for which $\mathfrak{S}^n M$ (for some non-negative n) has all its elements of the first level. Now, for the set Z of natural numbers belonging to a domain B, it is a set of the second level. If B' is the partial domain of sets of the first or second level, then all of the axioms hold also in B' but the collection formed by the elements $Z, \cup Z, \cup^2 Z, \ldots, \cup^n Z \ldots$ is not a *set* (i.e. is not an element of B').[5] Skolem's attempt to 'remove' this limitation is based on a new axiom:[6] "Let U be a definite preposition that holds for certain pairs (a, b) in the domain B; assume, further, that for every a there exists at most one b such that U is true. Then, as a ranges over the elements of a set Ma, b ranges over all elements of a set Mb". But, Skolem claims, this new axiom does not change anything towards the *relativity* of the axioms.

The third remark just proves that no domain B could be considered as the only one concerning set theory. By defining a *descending* $\in$*-sequence* of a set M: a sequence of the form $\ldots M_2 \in M_1 \in M$, where M, M_1, M_2, etc belong to the domain. If a domain B where the axioms hold is taken, those sets having only finite descending $\in$-sequence form a domain B' where the axioms also hold. If not all of the sets of B have finite descending $\in$-sequences, then already two different domains B and B' are shown. But if all the sets in B have only finite descending $\in$-sequences, the sets in B having their descending $\in$-sequences finishing with $\emptyset$, form a domain B' also valid for the set-theoretic axioms. Even it is possible to form a domain including also an object a not belonging to B; that is to say, adjoining a new element does not lead to a contradictory system. Skolem concludes "it would in any case be of much interest if we could prove that a new subset Z_0 could be adjoined without giving rise to contradictions".[7] One of Skolem's conclusions, considering the continuum power, says then:

> Since Zermelo's axioms do not uniquely determine the domain B, it is very improbable that all cardinality problems are decidable by means of these axioms. For example, it is quite probable that what is called the continuum problem, namely, the question whether $2^{\aleph_0}$ is greater than or equal to $\aleph_1$, is not solvable at all on this basis; nothing need to be decided about it.[8]

These three points concerning axiomatic set theory pointed out that from that moment any attempt to give a solution for the continuum problem within Zermelo's axioms would lead necessarily to a *relative* answer. Nevertheless, the question was still open whether an *absolute* answer could be given. These were the contents of Hilbert's attempt in 1925, and of Gödel's proof in 1939.

HILBERT'S METAMATHEMATICAL APPROACH

The central idea of Hilbert's work *On the Infinite* is that the *infinite* can only be treated and conceived within mathematics. That means, there is no possible theory of the infinite that would not be a mathematical theory.

And from the mathematical theories that in one way or another have dealt with the infinite, whether Euclidean geometry, infinitesimal analysis, or Cantorian set of theory, it is the latter which must be considered as the real and authentic mathematical theory of the infinite. With this thesis as point of depart, Hilbert states in his paper a double task: on the one hand to show how a mathematical theory of the infinite is possible without any contradiction – a theory where the question about the *possible* existence of mathematical entities would be over – and his second task, to show how this metamathematical thesis concerning the infinite could help to find a solution to the continuum problem, the most important mathematical problem concerning the infinite.

With Karl Weierstrass's revolution for mathematical analysis, all the imprecisions and difficulties derived from the obscure notion of *infinitesimals* were abandoned and substituted by precise and definite relations between finite magnitudes. But the problem of dealing with an infinite and complete totality in set theory has not yet been treated, says Hilbert, in a *finitarian Weierstrassian way*. Hilbert's proposal attains also some mathematical formulas, like those appearing in number theory, which involve infinitely many propositions.

The need of this new finitarian way to treat those infinite totalities comes out from the fact that they are mostly involved in logical and mathematical inferences, and in this sense they deal in general with mathematical thought.

Now, after the paradoxes that appeared in set theory some mathematicians thought that a theory for the infinite was but an illusion. Hilbert's point of view is that the problems come from the fact that logical inference needs, to be correctly applied in mathematics and in scientific knowledge, of some extra logical, *concrete* objects given in our intuition and "prior to all thought". This epistemological thesis, that Hilbert supports in Kantian philosophy, becomes his main philosophical thesis for mathematics: those concrete and extra logical objects for mathematics are nothing but *signs* themselves, "whose shape is immediately clear and recognizable". Sign intuition is taken by Hilbert as the ground where mathematics builds up its certainty. This is enough for finitary mathematics: logical inference applied to concrete signs could give us a formulation of any true proposition in finite arithmetic. But as soon as the infinite appears, this procedure does no longer hold: any existential proposition ("there exists a number with a certain property") could be considered, in the finite case, as a certain finite disjunction ("this, or that or one of this finite collection of numbers has this property"). But to say, for example, that there exists a number, greater than a certain fixed one, and having a certain property, is not reducible to a disjunction as in the previous case. Beyond this fact, Hilbert states the existence of two kinds of mathematical propositions: on the one hand, "concrete and finite statements" and on the other "statements without any sense and constituting the *ideal domain* of our theory". That means that propositions dealing with the infinite are to be considered as *ideal propositions*.

To deal with the truth value of a *concrete* proposition is nothing but to confirm whether or not it is the case what the proposition states. On the other hand the statement that a certain proposition is a true one is equivalent to the fact that its negation is false. But for the ideal propositions it is not possible, "in so far as they do not express finitary assertions", to consider logical operations in a contentful and concrete way. A purely formal treatment for logical procedures and to mathematical proofs is needed when dealing with these ideal propositions.

The only condition that these ideal propositions should hold is that their introduction should not contradict any already existing proposition. Contradictory propositions are not obtained from axioms *via* the formal scheme

$$\frac{\begin{array}{c}\mathfrak{S}\\ \mathfrak{S}\rightarrow\mathfrak{T}\end{array}}{\mathfrak{T}}$$

The finitary character of ordinary mathematics is kept as finitary procedure for transfinite mathematics. That is, in set theory – and the continuum problem is a major case – a finitary *proof theory* or *metamathematics* is due to guarantee the certainty that sign intuition does for finitary mathematics. All this leads Hilbert to a main metamathematical thesis: *every mathematical problem can be solved.* Out of this thesis, an affirmative answer for the continuum problem could be found (it is possible to 'count' the points of the straight line by means of the numbers of class II). Now, in order to proceed to his proof for the continuum hypothesis, Hilbert replaces the set of real numbers (or the points of a straight line) by an equivalent one: the set of number-theoretic functions. The way in which each function is generated is necessary to find a correct arrangement of this set so to define the correspondence with number class II. To avoid the problem represented by those functions whose definition depends upon the solution of some mathematical problems, Hilbert uses a restricted version of his metamathematical thesis about the solution of any mathematical problem. Stated as a metamathematical lemma, it allows him to restrict the solution only to those functions which can be defined by recursion:

> If a proof of a proposition contradicting the continuum theory is given in a formalized version with the aid of functions defined by means of the transfinite symbol $\in$, then in this proof these functions can always be replaced by functions defined, without the use of the symbol $\in$, by means merely of ordinary and transfinite recursion, so that the transfinite appears only in the guise of the universal quantifier: ( ).[9]

This means that if the cardinal number of the set of number theoretic functions is larger than the cardinal of number class II, then already the set of recursive integer functions is larger. That is why Hilbert states in a new way his problem: a biunivocal correspondence must be established between the numbers of the second class and those functions (with integer values and integer arguments) defined without the symbol $\in$.

But when dealing with these number theoretic functions, it is not possible to use only one simple recursive scheme for there are many different ways to pass from n to $n+1$. As an example of this impossibility, Hilbert shows the following recursive definition:

Starting with the function $a+b$, by n-fold iteration $a+a+\ldots+a = a \cdot n$.

In a similar way it is possible to pass from $a \cdot b$ to $a \cdot a \cdot a \cdot \ldots a = a^n$, and from a^b to $a^{a^{\cdots}} \ldots$

So the following functions have been defined:

$$\Phi_1(a,b) = a+b,$$

$$\Phi_2(a,b) = a \cdot b,$$

$$\Phi_3(a,b) = a^b.$$

$\Phi_4(a,b)$ being the bth term of the sequence a, a^a, $a^{a^{\cdots}} \ldots$ After that the functions $\Phi_5(a,b)$ and $\Phi_6(a,b)$ are defined. But the definition of the function $\Phi_n(a,b)$ might need not only ordinary recursion but manifold simultaneous recursion. The function $\Phi_a(a,a)$ is an example of such a function and it can be defined by ordinary recursion only if variable types are employed.[10] (Variable types are of type 1 if they are functions with integer argument and value; they are of type 2 if its value is a number and a function – a type 1 variable type – as an argument, etc.)

With these variable types a general form for recursion can be stated:

$$\begin{aligned} \rho(\mathfrak{g}, \mathfrak{a}, 0) &= a \\ \rho(\mathfrak{g}, \mathfrak{a}, n+1) &= \mathfrak{g}(\rho(\mathfrak{g}, \mathfrak{a}, n), n) \end{aligned}$$

with $\mathfrak{a}$ an expression with an arbitrary type; $\mathfrak{g}$ an expression with two arguments (the first one has the same type as $\mathfrak{a}$, the second one is a number variable); the value for $\mathfrak{g}$ has the same type as $\mathfrak{a}$; ρ is the defined expression.

Variable types are precisely those which allow Hilbert to define a correspondence between real numbers and the second number class. First a comparison is established between the generation procedure of these types according to their height, and the generation procedures of transfinite numbers of class II; to the adjunction of a unity it corresponds the substitution of a type as an argument in a higher type; to the limit of a sequence process corresponds the union of a countable sequence of types in a new type. Variable types so related with numbers of the second class are named Z-types. The generation of Z-types might be carried out only with ordinary recursion (besides the logical operations); in this way, by arranging these Z-types according to their heights, a correspondence is reached between the numbers of class II and Z-types. But it is also possible to reach a correspondence between number class II and those entire functions defined by means of Z-types. By giving an upper bound for the types employed in defining functions, only a countable infinite set of functions might be defined by recursion and substitution. On the other hand, the formal definition for the numbers of class II might be carried out only by ordinary recursion (in each limit process a new higher type is used so that ordinary recursion is needed only to enumerate the type sequences before this process). With these facts, Hilbert states:

> (...) we do not need to associate with a given number of the second class all functions of the same height, but we can let the numbers of the second number class and the functions correspond to each other according to the heights of the variable types required for the definition.[11]

With a certain height of type as a bound, a countable sequence of numbers of class II will be defined, and by the limit process a new number will be defined only by introducing a higher variable type. In the same way, the infinite countable set of functions defined by the types with the same height as a bound will allow to define a new function –

by th diagonal method – which could be defined only by higher types. Hilbert concludes:

> Thus we have made it possible to establish a one-to-one correspondence between the numbers of the second number class that are definable at the height in question, but not at any lower one, and the denumerably many functions that are definable at the same height, and in this way each function is associated with at least one number of the second number class.[12]

But the proof of the continuum theory by this method needs another assumption, namely, that ordinary recursion, as the one used for the correspondence, is enough to define all number theoretic function which can be defined recursively. This assumption is stated by a second metamathematical lemma:

> In the formation of functions of a number theoretic variable, transfinite recursions are dispensable; in particular, not only does ordinary recursion (that is, the one that proceeds on a number theoretic variable) suffice for the actual formation process of the functions, but also the substitutions call for those variable types whose definition requires only ordinary recursion.

So, based on two metamathematical lemmas, Hilbert considers that his proof theory has given a proof for the continuum problem. But two problems arise from Hilbert's attempt to prove definitely the correspondence between the two sets. On the one hand, his proof rests on those metamathematical lemmas which in turn were not proved. On the other hand, even if he insisted in the difference between the two 'mathematical regions': mathematics and metamathematics, his attempts makes this difference almost disappear. The problem, as originally stated by Cantor, is a *mathematical problem*; Hilbert insisted that its proof depends on a *metamathematical situation* (as expressed by lemma I), but from that moment the proof itself is a metamathematical one.[13] The two metamathematical lemmas state both the ideal conditions, according to the finitary formalist program, to give a solution to the continuum problem. The first lemma states that it should be solved without using any *choice* principle. The second lemma states in its turn that for the continuum, as well as for the numbers of class II, ordinary recursion is enough to dominate them.

The conditions assumed by those lemmas established that the continuum problem could never be solved without the complete solution of the metamathematical problem (understood as the statement of the situation of the problem). Following this lesson, but in a way that turned to be a return of the *relative conditions* announced by Skolem, was the way of K. Gödel's consistency proof.

GÖDEL'S CONSISTENCY PROOF

Gödel's proof that the generalized continuum hypothesis is relatively consistent with the Zermelo–Fraenkel axioms, seems to follow Skolem's lesson. He assures that the proof of consistency is relative to the axioms of set-theory by the construction of a *model* for this theory in which it can be checked that the hypothesis 'holds'. Indeed Gödel proves the consistency of the Generalized Continuum Hypothesis; this is the statement that for each ordinal number α, $2^{\aleph_\alpha} = \aleph_{\alpha+1}$ (Cantor's continuum hypothesis is the case $\alpha = 0$).

Gödel develops the model of *constructible sets*: for a given class M, where it is defined the $\in$-relation, an open formula over M is a formula which contains (i) variables, (ii) constants, (iii) the $\in$ sign, (iv) the negation sign, (v) disjunctive sign, (vi) the quantifiers for the variables M^* is the class of those subsets of M characterized by an open first-order formula $\Phi(x)$ on M (i.e. a subset M' of M will be an element of M^* if M' is the set of all those elements x of M such that $\Phi(x)$).[14] Then Gödel defines

$$M_0 = \{\emptyset\}$$

$$M_{\alpha+1} = (M_\alpha)^*$$

$$M_\partial = UM_\alpha \quad \text{(with } \alpha < \partial\text{) when } \partial \text{ is a limit ordinal.}$$

A set x is 'constructible' if there is an ordinal number α, such that $x \in M_\alpha$, and it is 'constructible of order α' if

$$x \in (M_{\alpha+1} - M_\alpha).$$

Under the assumption 'every set is constructible' the set $M\omega_\omega$ becomes a model for Zermelo's axioms, and M_Ω (with Ω the first inaccessible cardinal) is a model for all of Zermelo–Fraenkel axioms (i.e. including the replacement axiom). That means that in M_Ω – as considered by Skolem – all of the axioms hold, so that it can be conceived as a model for *constructible sets*.[15]

Now, for this model of constructible sets the following theorem is proved:

Any constructible subset m of $M\omega_\alpha$ has an order smaller than $\omega_{\alpha+1}$.

With this theorem it is now proved:

$$|M\omega_\alpha| = \aleph_\alpha \quad \text{for any ordinal number } \alpha.$$

If m is a subset of $M\omega_\alpha$, the theorem states that $m \in M\omega_{\alpha+1}$. That is, $\cup(M\omega_\alpha)$ (the set of all subsets of $M\omega_\alpha$) is a subset of $M\omega_{\alpha+1}$. This allows to conclude that $2^{\aleph_\alpha} \leq \aleph_{\alpha+1}$. From Cantor's theorem it is known that $\aleph_{\alpha+1} \leq 2^{\aleph_\alpha}$. These two results lead to the equality $2^{\aleph_\alpha} = \aleph_{\alpha+1}$.

Gödel states then that if A is the sentence "there are no non-constructible sets", and C is the sentence "for each ordinal α, $2^{\aleph_\alpha} = \aleph_{\alpha+1}$", then the statement $A \rightarrow C$ is a theorem.

Two main remarks should be pointed out considering Gödel's proof. Above all the fact that the consistency proof is *relative* to the consistency of the other axioms. This means that from that moment, the existence of a 'model for Zermelo–Fraenkel axioms' where the continuum hypothesis holds, allows us to conclude that if a contradiction ever appears in axiomatic set theory with the continuum hypothesis, this contradiction will appear also without the continuum hypothesis.

But a second remark has to do with the fact that the proof of the continuum hypothesis holds in the model of constructible sets. This fact sounds as a great restriction because only a special kind of set is taken into account; in other words, the theorem states that any non-countable *constructible set* of reals has the cardinal number of the continuum. But what about the non-constructible and non-countable sets in order to get a 'complete' proof for the continuum hypothesis? It should be pointed out that as in the case of Hilbert's approach to the problem, it is again a question about the possibility of attaining transfinity by means of constructive or recursive procedures. Indeed Hilbert's purpose for having a proof for the continuum hypothesis was precisely to have a certain equivalence of all the ordinal of number class II with the set of all recursive entire functions. Recursive functions would allow to define recursive or constructible sets; so Hilbert's second metamathematical lemma, assuring that ordinary recursion is enough in order to define all number theoretic functions defined recursively. This seems a statement close to Gödel's assertion that $\cup(M\omega_\alpha)$ is a subset of $M\omega_{\alpha+1}$ for the case when $\alpha = 0$. Gödel himself said that

> There is a remote analogy between Hilbert's Lemma II and my theorem for the case $\alpha = 0$. There is, however, this great difference that Hilbert considers only strictly constructive definitions and, moreover, transfinite iterations of the defining operations only up to constructive ordinals, while I admit not only quantifiers in the definitions but also iterations of the operations up to *any* ordinal number, no matter whether or how it can be defined. The term 'constructible set' in my proof, is justified only in a very weak sense and, in particular, only in the sense of 'relative to ordinal numbers', where the latter are subject to no conditions of constructivity. It was exactly by viewing the situation from this highly transfinite, set-theoretic point of view that in my approach the difficulties were overcome and a *relative* finitary consistency proof was obtained. Of course, there is no need in this approach for anything like Hilbert's Lemma I. Hilbert probably hoped to prove it as a special case of a general theorem to the effect that transfinite modes of inference applied to a constructively correct system of axioms lead to no inconsistency.[16]

So, as Gödel himself pointed out[17] a short time thereafter, a formal proof of consistency within axiomatic set theory, because of its *relative* sense does not give us an answer concerning Cantor's mathematical problem.

It is possible to distinguish two main problems concerning the power of the continuum, since the original formulation by Cantor, including his first hypothesis that the power of the continuum was that of the second number class, up to Gödel's consistency proof. On the one hand, the certainty that it is above all a mathematical problem which ought to be solved, and on the other hand, the fact that its solution goes beyond the possibilities given by axiomatic set theory.

The complete classification for powers of sets (at the beginning the problem concerned only point-sets) was a problem that Cantor tried to solve from the beginning of his theory. In his last writings, he kept the conviction that transfinite number theory (conceived in its twofold aspects of ordinal and cardinal numbers) was the correct and only way. But after Zermelo's axiomatization for set theory, the question arose whether by means of these axioms a complete answer to all problems concerning sets could be found.

Skolem's contribution was to state that within axiomatic set theory only *relative* answers could be given to many set theoretical questions, including the continuum problem. On the other hand, Hilbert tried to establish the metamathematical condition that could make it possible to solve the continuum problem. His metamathematical lemma asserting that any well-formulated problem should be solvable, pretended to give the definite condition, above the axiomatic limitations, to prove that the linear continuum could be numerated by the numbers of class II.

The twofold lesson of Gödel's proof is then to confirm once more the necessary relativeness of set theoretical notions when they are developed within an axiomatic system. Skolem and Hilbert's results are here retaken. But, even after that proof, Gödel's conviction that the *relative* answer just given by no means could solve the mathematical problem, leads us to find Cantor and Hilbert's conviction that the continuum problem is above all a mathematical problem. He even conjectured the possibility of finding the *independence* of the continuum hypothesis from the rest of Zermelo–Fraenkel axioms (given in 1963 by P. Cohen).

Two possible conclusions could be taken from what we have just said: whether the continuum problem is no longer a legitimate problem (after Gödel and Cohen's theorems); or the statement that besides Zermelo–Fraenkel axioms (or any other axiomatic system), set theory is something else.

NOTES

[1] D. Hilbert, 'Mathematical problems', lecture delivered before the International Congress of Mathematicians at Paris, 1900. Bulletin of the American Math Soc., Vol. 8, 1902. Translated by Mary Winston Newson.

[2] Although König insisted in disproving Cantor's Continuum Hypothesis by showing that the continuum could not be well ordered (his second attempt was made as lecture before the Hungarian Academy of Sciences in 1905), he never succeeded.

[3] Th. Skolem, 'Some remarks on axiomatized set theory'. English translation in J. van Heijenoort, *From Frege to Gödel.*

[4] $\mathfrak{S}^{n+1}M = \mathfrak{S}(\mathfrak{S}^n M)$ and $\mathfrak{S}M$ is the first union of the set M and $\mathfrak{S}^0 M = M$.

[5] No set of power $\aleph_\omega$ if the continuum hypothesis is assumed.

[6] In fact another formulation of the replacement axiom.

[7] Th. Skolem, *Ibid.*

[8] Th. Skolem, *Ibid.*

[9] D. Hilbert, 'Über das Unendliche'. English translation in J. van Heijenoort, *From*

Frege to Gödel.

[10] Ackerman proved in 1928 that this function grows faster than any ordinary recursive function of a (this means that for any recursive function $f(a)$, a number m can be given so that $\Phi_a(a, a) > f(a)$ whenever $a \geq m$.

[11] D. Hilbert, *Ibid.*

[12] D. Hilbert, *Ibid.*

[13] This was pointed out already by W. Sierpinski in the introduction to his book *L'Hypothèse du Continu* where he says that Hilbert's attempt is just "une esquisse de démonstration, liée encore avec des grandes difficultés, car plusieurs lemmes essentiels restent à établir".

[14] The sentence $\Phi(a)$ for an element $a \in M$ is equivalent, Gödel states, to the existence of Skolem functions for Φ and a. These are any n-adic and $(n + h)$-adic functions on M, $f_1, \ldots, f_m, g_1, \ldots, g_e$, such that for every $x_1, \ldots, x_n, z_1, \ldots, z_h, \ldots$ of M,

$$\Theta(a, x_1, \ldots, x_n, f_1(x_1, \ldots, x_n), \ldots, f_m(x_1, \ldots, x_n), z_1, \ldots, z_h,$$
$$g_1(x_1, \ldots, x_n, z_1, \ldots, z_h), \ldots, g_e(x_1, \ldots, x_n, z_1, \ldots, z_h), \ldots),$$

where the function Θ gives the normal form for $\Phi(a)$.

[15] M_Ω is a model for all of Zermelo–Fraenkel axioms; i.e. including the replacement axiom. For Zermelo's axioms $M\omega_\omega$ is already a model where the sentence "all sets are constructibles' is true.

[16] Letter of Gödel dated 8 July 1965, quoted by J. van Heijenoort in *From Frege to Gödel.*

[17] In this paper of 1946 'What is Cantor's Continuum Problem?' in Benacerraf, P. and Putnam, H., *Philosophy of Mathematics*, Princeton University Press.

MAX FERNANDEZ DE CASTRO

ON THE RELATION OF HILBERT'S SECOND AND TENTH PROBLEM

In August 1900, during the Second International Congress of Mathematicians, held in Paris, David Hilbert proposed 23 problems, well known since then. He tried to anticipate and, up to a certain point, to determine the subjects and predictable development of mathematics for the new century beginning. According to his opinion the vitality of a science in a given epoch depends directly on the number of problems raised, since they constitute the guide which directs the efforts of scientists.

Now, toward the end of the 20th century, it is important to review the status of research around every one of Hilbert's problems and to evaluate their degree of fecundity in the development of contemporary mathematics.

The present paper is related to two of these problems. The second, concerned with 'the non-contradiction of the axioms of arithmetic', and whose text was enunciated this way: (a) "To prove that they [the arithmetical axioms] are not contradictory, that is, that a finite number of logical steps based upon them can never lead to contradictory results" (Hilbert, 1900). And the tenth which refers to 'the possibility of solving a Diophantine equation': (b) "Given a Diophantine equation with an arbitrary number of unknown quantities and with rational integral numerical coefficients: To devise a process according to which it can be determined by a finite number of operations whether the equation is solvable in rational integers" (Hilbert, 1900).

Thirty years later the Austrian-born mathematician Kurt Gödel proved, in his famous 'incompleteness theorem', the impossibility of solving satisfactorily the problem mentioned in (a).

However, on the basis of the theory contained in the proof of that theorem, several authors, like S. Kleene, Martin Davis, Julia Robinson and others, made significant contributions toward the conceptual evolution of the tenth problem which was finally solved by Matiyasevic in 1970. The matter we deal with in the following pages is to sketch a proof, suggested by Julia Robinson and Martin Davis, relating both results and

S. Ramirez and R. S. Cohen (eds.), Mexican Studies in the History and Philosophy of Science, 187–199.

to comment briefly their importance in the history of Mathematics. With this aim we shall describe and analyze extensively Hilbert's program whose development was required to solve the second problem.

First, we must remember that during the 1900 conference, Hilbert expressed his conviction that any definite mathematical problem "must necessarily by susceptible of an exact settlement either in the form of an actual answer to the question asked, or by the proof of the impossibility of its solution and therewith the necessary failure of all attempts" (Hilbert, 1900). In both cases we discuss, the solution is of the latter type.

THE IMPORTANCE OF THE TENTH PROBLEM

A Diophantine equation is an equation in which both members are polynomials with any number of variables and with natural coefficients and exponents. For instance, $x^3 + 5xy = 2z^2$ and $x^n + y^n = z^n$ (with n a natural number) are Diophantine equations. When we speak of their roots we are referring to the natural roots and not to the integral ones, the reason being that the problem proposed by Hilbert is equivalent to that obtained by replacing this condition by the original. Its importance can be easily appreciated if one thinks that the well-known Fermat's problem ("there are no natural numbers x, y, z and n such that $x^{n+3} + y^{n+3} = z^{n+3}$") is a particular case of a Diophantine equation. Hilbert says that:

> The attempt to prove ... [Fermat's problem] offers a striking example of the inspiring effect that such a very special and apparently unimportant problem may have upon science. Kummer, incited by Fermat's problem, was led to the introduction of ideal numbers and to the discovery of the law of the unique decomposition of the numbers in a circular field into ideal prime factors – a law which today, in its generalization to any algebraic field by Dedekind and Kronecker, stands at the center of the modern theory of numbers and whose significance extends far beyond the boundaries of number theory into the realm of algebra and the theory of functions (Hilbert, 1900).

This passage shows the reasons that moved Hilbert to include the tenth problem among those he assumed to be of greater fruitfulness for the development of mathematical thought.

Initially the tenth problem consisted in finding a finitary procedure with the specified conditions, but the difficulties faced by those who tried to solve it directly, as well as the development of the theory of computability, shifted the solution towards the negative approach, i.e. that such a method did not exist. Moreover, the argument, which had been very ambiguous in 1900, is now precise if we accept Church's thesis which states that a proper characterization of the terms 'algorithm' or 'numerical calculus' is given by Turing machines.[1] In fact, the tenth problem refers to "a process through which ... by a finite number of operations" we can in a mechanical way solve certain types of problems.

Now, to prove that such a method does not exist, it is necessary to give a precise meaning to words. If one accepts Church's thesis, then a rigorous concept of what is a 'mechanical procedure' is derived from Turing machines or the λ-calculus. Obviously, Church's thesis cannot be proved, but there are many reasons for accepting it, for example, that different attempts to rigorously define the notion of computable function have been shown to be equivalent.

In order to frame correctly the importance of Hilbert's second problem, we shall expound (reduced to its minimal expression) what the formalist program claimed for the foundations of mathematics, taking into account that the central point of the program was precisely the above mentioned problem.

THE IMPORTANCE OF THE SECOND PROBLEM

The formalist program which pretended to solve the crisis in the foundations of Mathematics was postulated, each time more precisely, in a series of articles published by Hilbert during the 20's, especially those entitled 'On the Infinite' and 'The Foundations of Mathematics', upon which we shall concentrate our attention as they constitute the most complete expression of formalist position.

According to Hilbert, mathematical propositions are of two types: those that we could call 'real' propositions dealing with or extracted from objects of our sensible intuition, like the simplest arithmetical identities; and the propositions that refer to what he calls 'ideal notions'. For the latter, points to infinity in geometry and imaginary numbers are paradigmatical examples. These ideal notions are added to the body of mathematics with the purpose of completing it and in order to give them an elegance and symmetry that greatly facilitates the calculation or the postulation of general theorems. Thus, the line at infinity validates the duality principle of projective geometry, and – in the same manner – complex magnitudes were introduced in such a way "that the laws of algebra, those, for example, concerning the existence and number of roots of an equation could be preserved in their simplest form" (Hilbert, 1925).

One of these 'ideal notions', probably the most problematic in the history of mathematics, is that of infinity. It appears implicitly in the fifth postulate of Euclid's *Elements* and it is directly tackled, many centuries after, through the infinitesimal calculus where the concept reappeared in the form of infinitely small or 'evanescent' quantities. This produced very useful results by using very imprecise techniques but, eventually, brought confusion to mathematicians. The return to rigour was due, among others, to Karl Weierstrass who suppressed concepts wherein the notion of actual infinity was implicit, and substituted for them concepts and terms that, being equivalent, referred only to potential infinity.

Hilbert compared this program to deal with the paradoxes brought forth by the theory of sets with the efforts made by Weierstrass towards making analysis a rigorous discipline. One of Hilbert's proposals was to regard infinity as an ideal element. Antinomies, he says, emerge as the result of a misunderstanding of the way in which new notions should be introduced in a theory, and not by the unrestricted use of the principle of the excluded middle, as the intuitionists thought. In fact, real propositions are free of contradictions as long as they can be interpreted as sentences about objects given in intuition. Ideal notions, on the other hand, as they do not come from a direct source, should be introduced in a theory only with the proof that they do not generate contradiction. For this, in turn, it is necessary to formalize mathematics (as we shall see in the following section).

THE FORMALIST PROGRAM

For Hilbert, as for Kant, mathematics cannot be founded on logic alone. He states:

> ... as a condition for the use of logical inferences and the performance of logical operations, something must already be given to our faculty of representation; certain extralogical concrete objects that are intuitively present as immediate experience prior to all thought (Hilbert, 1925).

These objects, which must be totally given in intuition, are signs, whose form is immediately clear and recognizable.

Here, Hilbert finds himself following an evident tendency in the development of mathematics, i.e. the idea contained in Vieta's formula according to which algebra was the mathematics of numbers – *Logistica Numeralis* – expressed with figures (*species*), and transformed into *Logistica Speciosa*. While we remain in the sphere of the most elementary arithmetic, numerical equalities are useful for stating that two linguistic expressions have the same meaning: proof is based on content. But if we move to theory of numbers or to algebra, we note that the use of numerical variables allows us: (a) to express general propositions; and (b) to prove by following certain rules in which signs are taken now as the concrete objects of our intuition.

> Hence even elementary mathematics contains, first, formulas to which correspond contentual communications of finitary propositions . . . and which we may call the *real propositions* of the theory, and, second, formulas that – just like the numerals of contentual number theory – in themselves mean nothing, but are merely things that are governed by our rules and must be regarded as the *ideal objects* of the theory (Hilbert, 1927).

Hilbert's proposal is to extend this algebraic method to the whole of mathematics and to regard the latter, at least for foundational purposes, as a science of signs, not of numbers or magnitudes. His plan consisted, first, in completely pouring mathematical content (in particular, arithmetic) into formalized axiomatic systems where a specification was made about the signs used, their syntax, axioms, and the rules for the derivation of theorems. Moreover, Hilbert claims that the consistency of axioms in a theory of this type is equivalent to the existence of the objects implicitly defined by them.

Furthermore, since a formal system is made of signs of which we clearly have an intuition, it should be possible to prove consistency by exclusively finitary and constructive methods. In order to do it, it is necessary to specify precisely the formal structure and the vocabulary of the system for which a consistency proof is sought. The system should be regarded as a body of finite expressions built in accordance with a series of formation-rules and a series of symbols that have been previously determined. Thus the ways of argumentation used in mathematics will be transformed into something objective that can be rigorously analysed and confirmed. Hilbert thus inaugurates a discipline directed to the study of formalized mathematics, and toward the investigation of the properties of such systems. He calls it *Metamathematics* or *Theory of Proof*. Since it is to provide the consistency of a huge range of theories, and particularly those of elementary arithmetic, its procedures cannot go beyond the sphere of the finitary. It is thus that Hilbert copes with intuitionist criticism.

THE FINITIST CRITERION

We recall that, for Brouwer and his group, the existence of an object must be established by constructive methods, i.e., it is only achieved through the exhibition of procedures that produce the object in a finite time. Thus, for example, the Euclidean theorem that asserts the existence of an infinite number of prime numbers, is acceptable for it states that, given the nth number of that series $p(n)$ there is a greater one smaller than $p(n)! + 1$ (Hilbert, 1925). Hilbert uses this example to illustrate how the intuitionistic position would lead us to abandon Aristotelian logic. Such a position is untenable: from a proposition which states that between two terms of the numerical series there is one that is prime, a proposition is inferred asserting the existence of an element in an infinite series. This consequence is not valid from the intuitionistic point of view because the negation of that sentence cannot be finitarily verified.

Hilbert advances his program to solve these problems:

> The fundamental difficulty that we face here can be avoided by the use of ideal propositions. For, if to real propositions we adjoin ideal ones, we obtain a system of propositions in which all the simple rules of Aristotelian logic hold and all the usual methods of mathematical inference are valid (Hilbert, 1927).

Yet, the consistency proof that validates the addition of the ideal propositions has to be carried out through finitary procedures based on the direct intuition of signs.

Thus Hilbert rejects the intuitionistic criterion for the validity of a proof in the mathematical field, while accepting it in his theory of proof.

After the process of arithmetization of analysis, carried out in the past century, and because a great number of consistency proofs for important branches of mathematics assumed the consistency of arithmetic, it became necessary to prove the consistency of arithmetic in a finitary and absolute way. The formalist program seemed to provide the means for doing so and to escape from the crisis of foundations.

In the same manner by which a finitary or constructive proof for the existence of a number provides the means to find it, algorithms deterministically preestablish an activity to obtain the results of an operation. In both cases, the work to be done requires a finite, yet indefinitely long time, in such a way that nothing is left to chance or to the operator's ingenuity. With all this we try to suggest how both Hilbert's problems that we are dealing with are closely connected, even if this connection is not immediately apparent.

By 1930, Hilbert's program was fully launched and it was the only direction that looked sound. However, a year later, an article by Kurt Gödel cancelled all hopes by proving two disconcerting results: one that establishes that any formal theory (F) that is consistent and suitable enough to include number theory (in which recursive functions and relations are representable and the axioms of arithmetic are either axioms or theorems), contains undecidable sentences or – equivalently – elementary arithmetic is not transferable, without loss, unto a formal axiomatic system. That is the first Gödel result. The second one, which appears as a mere corollary, asserts that either the consistency of F cannot be proved within F, or that it would have to be established using notions and methods that cannot be formalized in F. Gödel's proof relies on the most strict finitary and constructivist criteria, as was pointed out by Brouwer and his school.

Before sketching a proof of Gödel's theorem, taking as point of departure the solution of the tenth problem, we shall clarify some concepts related to the latter that are necessary to understand the proof we are going to provide.[2] We shall start with the concepts of Diophantine predicates and recursively enumerable sets.

DIOPHANTINE PREDICATES

Definition. A subset A of N^m $(m > 0)$ is *Diophantine* if there is a polynomial $P(x_1, \ldots, x_m, y_1, \ldots, y_k)$ such that

$$A = \{(x_1, \ldots, x_m) \mid E(y_1, \ldots, y_k)P(x_1, \ldots, x_m, y_1, \ldots, y_k) = 0\}$$

(where we have written '$E(y_1, \ldots, y_k)$' instead of '$(Ey_1) \ldots (Ey_n)$').

We shall also call the predicate Diophantine $R(x_1, \ldots, x_m)$ defined by

$$R(x_1, \ldots, x_m) \leftrightarrow E(y_1, \ldots, y_k)P(x_1, \ldots, x_m, y_1, \ldots, y_k) = 0.$$

(Henceforth we shall write $X^{(m)}$, $Y^{(n)}$, etc., instead of $(x_1, \ldots, x_m)$ and $(y_1, \ldots y_n)$.

Example: The set B of the square numbers and its complement

$$x \in B \quad \leftrightarrow \quad (Ey)(x = y^2)$$

$$x \in B' \quad \leftrightarrow \quad (Ew)(Ey)(w^2 + 1 + y = x) \quad \text{and}$$

$$(Ez)(x + 1 + z = (w + 1)^2)$$

can be described in only one equation:

$$x \in B' \quad \leftrightarrow \quad (Ew)(Ey)(Ez)[(w^2 + 1 + y - x)^2$$

$$+ (x + 1 + z - (w + 1)^2)^2 = 0].$$

In this case we have used a very general technique that allows us to define a Diophantine set by means of a system of simultaneous equations $P_1 = 0$, $P_2 = 0$, $\ldots$, $P_n = 0$, and replacing them by the equivalent equation $P_1^2 + P_2^2 + \cdots P_n^2 = 0$. Similarly, a Diophantine set can be defined through a predicate of the type $P_1 = 0$ or $P_2 = 0$ or, $\ldots$, or $P_n = 0$, replacing it by the equation $P_1 \cdot P_2 \cdot \ldots P_n = 0$.

Moreover, the class of Diophantine predicates is closed under the logic operations of conjunction, disjunction and, of course, of existential quantification. On the other hand, negation allows us to attain, sometimes but not always, a Diophantine set from another one. The joint efforts by Martin Davis, Julia Robinson and Matiyasevic showed that such a class is also closed under bounded universal quantification. In fact, they proved that a set is Diophantine if and only if it is recursively enumerable.

Definition. A subset B of N is *recursively enumerable* either if $B = \emptyset$ or if B is in the range of a primitive recursive function.

The notion of recursive enumerability can have many different formulations and it is easy to generalize it in more dimensions. For our present aims it is enough to deal with the given definition. We remark that it is a simple outcome of computability theory that a Diophantine set is recursively enumerable. We shall summarize in five steps the proof of the inverse theorem, i.e. that from a recursively enumerable Diophantine set, the negative solution of tenth problem derives (because we know the existence of a recursively enumerable set which is not recursive):

1. (Gödel, 1931) Every relation $\{(X^{(n)}, y)I\ y = f(X^{(n)})$ where f is a recursive function can be defined by a Diophantine equation preceded by a finite number of existential and bounded universal quantifiers.
2. (Davis, 1953) Given a recursively enumerable set S, there exists a polynomial $P(x, y, u, x_1, \ldots, x_n)$ such that

$$\begin{aligned} S &= \{x \mid (Ey)(u)_{u \le y}(Ex_1 \le y) \ldots (Ex_n \le y) \\ & \quad P(x, y, u, x_1, \ldots, x_n) = 0\}. \end{aligned}$$

   The predicate defining S is called 'Davis normal form'.
3. (Davis, Putnam, Robinson, 1961) If the relation $z = x^y$ is Diophantine, every recursively enumerable set is Diophantine.
4. (Robinson, 1952) The relation $z = x^y$ is Diophantine if there exists a Diophantine set D such that:
   (a) $(u, v) \in D$ implies $v \le u^u$
   (b) For each k, there is $(u, v) \in D$ such that $v > u^k$.

   The antecedent of this implication is known as 'the Julia Robinson hypothesis'.
5. In 1971, Matiyasevic verified the Julia Robinson hypothesis. For instance, the set $D = \{(u, v) \mid v = s_u \& u > 3\}$ where

$$\begin{aligned} s_0 &= 1 \\ s_1 &= 2 \\ s_{n+1} &= 4s_n - s_{n-1} \end{aligned}$$

   is Diophantine and it satisfies (a) and (b).

From the latter point we conclude the solution of the tenth problem which, despite its negative aspect, shows an interesting equivalence between the concept of a Diophantine set and a recursively enumerable set. It opens new roads in the search for a decidable class of Diophantine equations.

GÖDEL'S THEOREM

The modality of Gödel's Theorem that we shall present here, establishes something a little more specific than the original theorem. We shall show that for any formal and consistent axiomatization of elementary arithmetic there is a Diophantine equation that does not have natural roots, "... but for which there is no proof from the given axioms that it is unsolvable" (Robinson, 1969).

The limited extension of this paper constrains us to sketch the central lines that constitute the proof. Let F be a theory of first order that satisfies the following conditions:

(a) F has among its symbols one that corresponds to each natural number (denoted by n', if $n \in N$), and another three ($+'$, $\cdot'$ and $='$) that under standard interpretation represent addition, multiplication and equality.

(b) For any three natural numbers n, m and r:
1. If $n = m$, then $\vdash_F n' =' m'$ and if $n \neq m$, $\vdash_F \neg n' =' m'$ (where '$\neg$' is the symbol for negation).
2. If $n+m = r$, then $\vdash_F n'+m' =' r'$ and if $n \cdot m = r$, $\vdash_F n' \cdot' m' =' r'$.

Hence, for any two polynomials $P(X^{(n)})$ and $Q(X^{(m)})$ $(n, m \in N)$ with natural coefficients there are two linguistic expressions $P'(X^{(n)})$ and $Q'(X^{(m)})$ in F such that for any sequence of natural numbers a_1, ... and a_m (assuming $n < m$) if $P(a_1, \ldots, a_n) = Q(a_1, \ldots, a_m)$ then

$$\vdash_F P'(a'_1, \ldots, a'_n) =' Q'(a'_1, \ldots, a'_m).$$

Expressions of the type $P'(X^{(n)}) =' Q'(X^{(m)})$ will be called polynomial predicates. Thus, for instance, the polynomial predicate $3'X_0 \cdot X_0 +' 4'X_1 =' X_3$ corresponds in F, to the equation $3X_0^2 + 4X_1 = X_3$. There is a one-to-one function g associating a number to each one of the symbols, expressions and sequences of expressions in F. Furthermore there exists an effective procedure for determining, given any natural number, if it belongs to the range of g, and, if so, we can also determine that element of the domain whence it comes from. Furthermore, a method exists for finding the image of any expression or symbol. Here, $g(u)$ means the Gödel number of u, and if $g(u) = n$, then we shall write $\exp(n) = u$. For each metamathematical relation or property R between the elements of our formal system, there corresponds, through g, an arithmetical relation $R^\sim$ between their respective Gödel numbers.

For instance if $R(x_1)$ is a metamathematical predicate of formulas, like x_1 being axiom, then let

$$R^{\sim} = \{n \in N \mid R(\exp(n))\}.$$

4. For F, the arithmetical relations that correspond to the three metamathematical predicates defined as follows, are Diophantine:[3]

(a) $FMAT(x) \leftrightarrow x$ is a polynomial predicate.
(b) $PR(y, w) \leftrightarrow w$ is a proof, in F, for the negation of y.
(c) Subst$(w, z) \leftrightarrow w$ is a formula and it is obtained from z by substituting in z the variable x_0 by n' where $n = g(z)$.

We shall enumerate now the polynomial predicates in the following way. If n is the Gödel number of a polynomial predicate, then it will occupy the nth place in our list. In a different case, let us make $0' = 0'$ the nth predicate. We shall denote by $Pn'(n', x_1, \ldots, x_m) = Qn'(n', x_1, \ldots, x_m)$ the nth term in this order.

The set

$$D = \{n \in N \mid \vdash_F \neg Pn'(n', x_1, \ldots, x_m) = Qn'(n', x_1, \ldots, x_m)\}$$

is Diophantine, since

$$n \in D \leftrightarrow FMAT(n) \& (Ew)(Ey) PR(w, y) \& \text{Subst}(w, n).$$

As D is a Diophantine set, there is a polynomial $P_j(x_0, \ldots, x_n) = Q_j(x_0, \ldots, x_n)$ such that

$$n \in D \leftrightarrow E(a_1, \ldots, a_n) P_j(n, a_1, \ldots, a_n) = Q_j(n, a_1, \ldots, a_n).$$

Moreover, if $n \in D$ then

$$\vdash_F P_j'(n', a_1', \ldots, a_n') = Q_j'(n', a_1', \ldots, a_n')$$

(for certain $a_1, \ldots, a_n$ natural numbers).

Theorem. If F is consistent, then $J \in' D$, i.e. the equation $P_j(j, x_1, \ldots, x_n) = Q_j(j, x_1, \ldots, x_n)$ has no solution, but this can not be proved in F.

Proof. If J belongs to D, there would be $a_1, \ldots, a_n$ natural numbers such that $\vdash_F P_j'(j', a_1', \ldots, a_n') = Q_j'(j', a_1, \ldots, a_n')$ and, therefore,

$$\vdash_F E(x_1, \ldots, x_n) P_j'(j', x_1, \ldots, x_n) = Q_j'(j', x_1, \ldots, x_n).$$

But, on the other hand, because of the definition of D, we would have that:

$$\vdash_F \neg P_j'(j', x_1, \ldots, x_n) = Q_j'(j', x_1, \ldots, x_n)$$

and, therefore,

$$\vdash_F (x_1)\ldots(x_n)\neg P_j'(j', x_1, \ldots, x_n) = Q_j'(j', x_1, \ldots, x_n)$$

contradicting the consistency of F.

Similarly, it can be shown that, if F is w-consistent, the enunciate $(x_1)\ldots(x_n)\neg P_j'(j', x_1, \ldots, x_n) = Q_j'(j', x_1, \ldots, x_n)$ is undecidable.

CONCLUSIONS

The connection between these two problems is not unexpected if we think about the close relation of the finitary criterion necessary for metamathematics and the notion of computability, which was vaguely sketched in the formulation of the tenth problem. That connection was not noticed by Hilbert who, nevertheless, foresaw, to a great extent, the course of mathematics in our century. The solution of the problem of Diophantine equations required, on the one hand, several techniques and concepts due to Gödel, as the characterization of finitary arithmetic by means of recursiveness and, on the other hand, the development of computability theory and the very detailed study of certain kinds of the Pell equation.

The most important lesson we may obtain from this matter is to notice how the enlightenment achieved in the intuition of a concept allows us to see the link between two problems that Hilbert could only present as isolated, and to suspect an underlying rationality for mathematics.

NOTES

[1] Which is equivalent to say that they are given by the λ-calculus or by recursive functions.

[2] Proof suggested by Julia Robinson (Robinson, 1969).

[3] Or recursive, because if the predicate $P(x, y)$ is recursive, then $(Ey)P(x, y)$ is Diophantine.

REFERENCES

Davis, Martin, 'Arithmetical problems and recursively enumerable predicates', *J. Symbolic Logic* **18**, 1953.

Davis, Martin, *Computability and Unsolvability*, McGraw Hill, New York, 1958.

Davis, Martin, Hilary Putnam and Julia Robinson, 'The decision problem for exponential Diophantine equations', *Ann. Math.* **74**, 1961.

Davis, Martin, 'An explicit Diophantine definition of the exponential function', *Comm. Pure Appl. Math.* **24**, 1971.

Hilbert, David, 'Mathematische Probleme'. Vortrag, gehalten auf dem internationalen Mathematiker-Kongress zu Paris', *Göttinger Nachrichten*, 1900. (English translation in *Proceedings of Symposia in Pure Mathematics* **28**, 1976.)

Hilbert, David, 'Über das Unendliche', *Mathematische Annalen* **95**, 1925. (English translation in Van Heijenoort, *From Frege to Gödel*, 3rd ed., Harvard, 1977.)

Hilbert, David, 'Die Grundlagen der Mathematik', *Hamburger Mathematische Einzelschriften* **5**, Teubner, Leipzig. (English translation in Van Heijenoort, *Ibid.*)

Gödel, Kurt, 'Über Formal unentscheidbare Sätze der Principia Mathematica und verwandter Systeme', *Monatshefte für Mathematik und Physik* **38**, 1931. (English translation in Van Heijenoort, *Ibid.*)

Robinson, Julia, 'Diophantine decision problems', in *M.A.A. Studies in Mathematics, Vol. VI: Studies in Number Theory*, W. J. Lebeque (ed.), 1969, pp. 76–116.

Rogers, H., *Theory of Recursive Functions and Effective Computability*, McGraw Hill, New York, 1967.

SANTIAGO RAMIREZ CASTAÑEDA

THREE METAPHYSICAL THESES ON MATHEMATICAL PHILOSOPHY*

The main purpose of this paper is to consider some of the ideas studied in the course that Jean Cavaillès taught at the University of Paris in 1941. In order to do this, I will use the manuscripts of notes taken by Mme. Marie Louise Gouhier and Mme. Geneviève Rodis-Lewis.

Jean Cavaillès taught but a single course at the Sorbonne. It was in 1941, when he took the place of Poirier who was sent on a mission to Brazil. At that time he was at Clermont-Ferrand, writing with Emmanuel D'Astier de La Vigerie and Georges Canguilhem the first leaflet of what was to be the group *Libération* and offering his own post there to Canguilhem.

According to the testimony of his sister, Gabrielle Ferrières, he accepted the post at Paris in order to be more effective in his clandestine activities.

Six months later, he was arrested at Narbonne. On December 29, 1941, he escaped from the camp at Saint-Paul d'Eyjaux "having with him the manuscript of his treatise on logic". This manuscript is lost. Still he had "fortunately given a typed copy to Mme. Foulquier ... this copy carefully kept by Mme. Foulquier but, helas!, incomplete, served for the publication of *Sur la logique et la théorie de la science* ..."

In 1941, Marie-Louise Defour and Geneviève Rodis attended Cavaillès's course. For the next forty-three years, they kept the notes they had taken. It is from fragments of these that I present the following paper. The text of Mme. Gouhier consists of eighteen pages (recto, verso); the last page is a summary of the themes about which the students were to speak. The text is divided in three parts (*'Relatif/absolu'*, *'Causalité'* and *'La notion de probabilité'*). The fragments over which I have worked come from the first part.

The second set of notes, those of Mme. Rodis-Lewis, consists of two thick notebooks. The first one has been unbound and it is Mme. Rodis who has tried to return it to its original order. It is made up of thirteen pages of a very small handwriting, full of abbreviations. The second set

S. Ramirez and R. S. Cohen (eds.), Mexican Studies in the History and Philosophy of Science, 201–211.

of her notebooks was kept properly. It is made up of thirty-three pages under the title '*Course de mardi*'. It is from the first set, rearranged to follow Gouhier's order, that I speak of today.

I

1. *The aim of science*, according to Cavaillès, *is to attain the absolute.*
We can look upon this proposition from two points of view: first from a Kantian perspective which would be related to Wittgenstein's *Tractatus*[1] and, second, to the theological works that Cavaillès had written several years before.

1.1. From the first point of view, I would like to emphasize the fact that Cavaillès looked upon Kant's *Critique of Pure Reason* with the same outlook that Heidegger held;[2] that is, he read it as an ontological text against the epistemological interpretations held by Cassirer.

From this point of view and in relation to mathematics, Cavaillès claims that the necessary outlook must be the ontological perspective. If we were to add Wittgenstein's influence to the above which induced Cavaillès to claim that logic is necessarily transcendental,[3] we must conclude that mathematics has nothing to do with the world because it is determined by empirical sciences: Mathematics aims to supersede the limits of experience.

1.2. From a theoretical point of view, Cavaillès stated, following Romano Guardini, his belief about the consummation of modernity. That is to say that if we recall that modernity commences with the construction of an infinite subject and with the acceptance of a geometrical representation of the world, contemporary mathematics is a restoration of the finitude of subjectivity and represents the end of a geometrical world. This claim is not a philosophical one but rather, as Cavaillès reiterates, a principle brought forth by mathematics itself. Thus, the historical pretense through which the essence of mathematics would be found in its relation with the empirical world is unsound. One must, rather, write the history of the efforts to restore to mathematics its own domain and one must write the history of the mirage, the illusion, that the mathematical realm belongs to this world.

2. The second claim which I would like to pose contends that *the history of the philosophy of mathematics sways between the Cartesian demand for the absolute and Kant's 'hypocrite'*[4] *absolutism.*

2.1. On one hand, Cavaillès states that Descartes illustrates the position of an absolute, infinity, which would be responsible for the internal unity of every mathematical notion and which would rationalize and authenticate them. Nevertheless, following Cavaillès, it is impossible to make precise the relation between the infinite and the finite. How, he asks, could one link the immutability of the object with the 'radical simplicity' of the infinitude of space such as it is grasped by the evidence?

The Cartesian solution consists in the introduction of an order that would explain the transition from simple intuitions to complicated systems as necessary, and such a necessity is to function as the guarantee of an absolute development of knowledge.

Here we face the arbitrary: the arbitrariness of a separation is a factor of relativism; the separation, at any rate, is not possible but for certain very particular cases. Thus, when the distinction between mechanical and geometrical curves is made, something belonging to space is lost.

2.2. On the other hand, the Kantian solution advocates absolute space as a condition for all knowledge. This argument has a point of departure, evidence to point to objects. Hence, concepts are not independent of their construction.

2.3. It is from these two possibilities that the development of mathematics may be made intelligible. Hence, in his *Logique et la théorie de la science*, Cavaillès explains a twofold movement of mathematics: the one inaugurated by Bolzano which would terminate with Husserl and the one following the path from Brunschvicg to Brouwer.

However, the absolute would not be attained in either way. One could, naturally, foresee the failure of the mathematical 'schools' of the 20th century, not as a deficiency of mathematics but as a failure in the method with which philosophy has worked within mathematics.

3. My third claim is that *mathematics is the science of infinity.*

First of all, Cavaillès says that the absolute to which mathematics aims is infinity. In the Kantian tradition, infinity is interpreted as a feature of space and time, but this infinitude of space would imply that only one geometry is possible.

In such an interpretation, mathematical knowledge would be reduced to an instrument for physical knowledge and, if so, there would be a "destruction of the value of mathematical knowledge: it is nothing more than the condition for the experimental knowledge of things".[5]

Hence, mathematical thought is only one way of thinking of the world. "Does it have a sense?" And, soon after, in his *Logique* he wrote: "even the Euclidean model cannot be preserved".[6]

In fact, if one is to observe that mathematics is the science of infinity, one is rejecting Kantianism. The only solution to retain infinity in mathematics is to be sought in the development of mathematics before any definition of its object.

It is thus that one can understand, both, the attempts to reduce mathematics to logic and to set theory.

The introduction of the infinite to mathematics brings out the paradoxes which will make logicism impossible.

We are led towards Hilbertian metamathematics. This means that, inasmuch as mathematics is held to be a 'gigantic formalism', there will be no contradictions: given an axiomatic system one can always define an object through them.

One concludes, with Cavaillès, that "only the infinity of natural numbers is absolute, all the rest are relative", and moreover, "every time that a system of objects is fixed by a set of axioms, the objects, as well as their properties, appear as relative to such a set. One must possess an absolute beforehand, a set from which one could choose".

If one were to give another interpretation of this third claim, one would have to place the problem in a different perspective which would have to take into account the following points:

(a) Mathematical infinity is only an abbreviation.
(b) The examples which one could give of it point always beyond that which is given.
(c) This indication is a finite suggestion, hence it belongs to this world. Still it points elsewhere. Mathematical thought is to be found in this world but it is always pointing beyond.
(d) The above means that when mathematicians think about this world, they will always give finite examples which aim farther, beyond every possible experience. Mathematics is not interested in objects of experience except in the measure in which they can be thought of otherwise.
(e) In this sense, mathematical ideas belong to the world but will not yield to the limits of language or of experience. They suggest transcendentality, that is, they address that about which one cannot have an experience and that about which one cannot speak.

(f) Regarding the history of mathematics, one cannot think of it as the history of a kind of thought concerning that which we can experience. Rather, it is the history of a kind of thought which has no history. Thus, Cavaillès can say that "the history of mathematics is not a history".

II

If we assume the above, we could state what I consider to be the three main issues of mathematical philosophy; namely,

1. The relation between mathematics and philosophy;
2. The relation between mathematics and experience; and
3. The relation between mathematics and reality.

We shall follow three philosophical inklings that we adopt as suggestions:

(a) Wittgenstein's *Tractatus* 6.54;
(b) Heidegger's construing of Kant's *Critique*; and
(c) Lecourt's way of envisaging philosophy as he proposes in *Une philosophie sans feinte*, that is, philosophy can no longer be considered as an enterprise towards unification; it must be deprived of the power that it has wielded over every other kind of discourse.

II.1

(a) Adopting Wittgenstein's 6.54 we must then regard philosophy as something that must replace (*überwinden*) its own propositions in order to arrive at an accurate vision of the world.

We can explicate Wittgenstein's proposition by stating that, regarding philosophy and mathematics, philosophy can only sketch a domain of being and that this task is over as soon as 'being' is susceptible to a further mathematical determination. Thus, the understanding of Wittgenstein's philosophical proposition coincides with the development of mathematics which determines the being philosophically sketched as a strictly mathematical being.

These determinations (which were traditionally grouped under the category of quantity) are autonomous in their relation with other possible descriptions of being. I will use the term 'weak autonomy' to describe this property.[7]

As an example of the consequences of 'weak autonomy', we can recall the Cartesian delusion according to which mathematical knowledge (being autonomous in the sense I have just specified) could become the norm to which every possible knowledge had to conform. We can also remember Comte's dream of transforming mathematics into the foundation of every positive science or, last but not least, Carnap's project of a perfect translation, of unified science, or of a complete and consistent axiomatisation.[8]

Still, my claim must not only relate to the place that mathematical knowledge takes but also, it must look at the position in which philosophy is now placed. To further my claim I bring up the second suggestion:

(b) In 1929, against Cassirer, Heidegger proposed an intriguing interpretation of Kant's *Critique of Pure Reason.*[9]

In one of his central points, Heidegger claims that Kant's *Critique* is the explicit foundation of metaphysics. That is, Kant's *Critique* does not deal with a theory of knowledge and, in particular, it does not deal with a theory of mathematical knowledge. In the lectures that Heidegger delivered that same year he clarified the precise sense of this contention as follows:

Pre-Kantian metaphysics, he said, specially Christian metaphysics, contemplated a partition of the totality of beings into God, Man and Nature and the corresponding division of metaphysics into theology, psychology and cosmogony. These three together would make up *metaphysica specialis.* "Being in general" would be the subject-matter of *metaphysica generalis.*

Within this conception, and inasmuch as the metaphysical object was something about which 'everybody is interested', metaphysics became the 'queen of the sciences'; in the same measure, the mode of metaphysical knowledge was to adjust itself to the ideal of mathematical knowledge, to the perfection of "pure rational science" which was to be autonomous of any possible experience.

Between this pure rational knowledge and knowledge of the particular regions of being there is, according to Heidegger, a trespassing that can, in particular, go beyond the sensible. It is, says Heidegger, a 'manner of behaviour' before, in front of, being.

Kant, indeed, stated that nature philosophers "learned that reason has insight only into that which it produces after a plan of its own and that (they proceed by) constraining nature to give answers to questions of reason's own determining".[10] Thus, we deal with a preconceived plan which assumes a manner of behaviour in front of being to which every possible research is referred.

Kant's claim suggests, at first glance, that natural philosophy by overstating mathematics' (weak) autonomy, lead to the notion that natural philosophy had to make permanent references to a being that had been conceived elsewhere and according to a previous plan.

If this plan, as pre-Kantian metaphysics leads us to believe, was a mathematical plan, if it is presented as a mathematical formalism whose object is being in general, whose forms are all inferred *a priori* and are continually new, mathematics is already an ontology.

Yet it is Kant himself who shows in the 'Transcendental Doctrine of Method',[11] that mathematics cannot be a *metaphysica generalis*. In this sense, mathematics owes to philosophy the shallow project of being, and mathematics cannot take over the task of becoming that determination of being to which every other research will be referred. In this sense, every other research demands for itself the same autonomy that mathematics has. I call this peculiarity (weak autonomy for all other research) strong autonomy.

In order to clarify the notion, I shall present an example which is central to the understanding of strong autonomy. In a well-known paper, Hilbert proves that the infinite, as determined by mathematics, owes nothing to the marks that astronomy or physics stamp upon it. Mathematical infinity, even if suggested by philosophy, can exist only by virtue of strictly mathematical criteria, that is, it can exist only because of the fact that its introduction creates no inconsistencies. Thus, a being, determined in a mathematical manner, owes nothing to other types of determinations and is rigorously autonomous (weak sense).[12] On the other hand, if we claim that this determination of the infinite determines the concept of infinity in other areas, that is equivalent to the claim that mathematics can serve as the foundation of every other possible research. That is to assume that the regions of particular being are mathematical models, that the particular regions of being are interpretations which make a formal system true. Mathematically one can prove that this is not the case.

(c) We can now look at the third suggestion. Our task is to contest every claim that would transform philosophical discourse into a discourse of legitimation and that would transform the discourse of philosophy into a discourse of unification.

In this sense, at least in its relation to mathematics, it is possible to show that philosophy can no longer have jurisdiction over mathematics and that mathematics works as a constraint against any such pretense, that is, mathematics functions as a critique of philosophy. This critique is a positive one inasmuch as mathematics exhibits that which philosophy can no longer think about. Mathematics bridges the gap beyond philosophy and points towards transcendentality.

Secondly, as mathematics demands an exclusive role in the task of the mathematical determination of being, as it yields to other determinations of being and as it recognizes in the latter the same autonomy it claims for itself, being must be thought of in its multiplicity; philosophy would be, thus, the act of thinking of being as a scattered being.

II.2

In reference to the relation between mathematics and experience I claim that this is strictly contingent.

In order to sustain my claim I return once more to Kant.

It was at Davos, again, that Heidegger stated the finitude of human reason, the finitude of the scope of sensibility and claimed that thought is finite "on its four sides"; furthermore, the need and the presence of thought confirm such finitude according to Heidegger.

Finally, then, we deal with the finitude of subjectivity, the finitude of intuition and the finitude of thought which is characterized by the noncreative trait of intuition: "what intuition must present in its singularity must be evident from the start".[13] Such finitude is the basis for experience that, as such, will not be capable of trespassing its own essence.

According to Heidegger, the understanding of being is brought forth among other beings: both the being which he is and the being which he is not are conspicuous to man. We call such a form of existence 'the being of man'.

> Man, in the manner in which he behaves in the presence of that being which he is not, discovers being as that which supports him and as that which he will never master . . . [14]

Such an existence is finite.

Yet, the essence of mathematics shows itself as infinity and its content as something radically and permanently open; there is no possibility of a mathematical experience, or of a finite experience of the infinite being of the 'actual' mathematical object. In any case, one can show finite examples. But what mathematics shows is not meant to be applied to finite examples. Mathematics points beyond.

Kant himself was conscious of this fact when he wrote:

> The latter [non-empirical intuition – S.R.] must, as intuition, be a single object, and yet nonetheless, as the construction of a concept (a universal representation), it must in its representation express universal validity for all possible intuitions which fall under the same concept (. . .) *mathematical knowledge* [*considers,*] *the universal in the particular* . . . [15]

In Wittgensteinian terms, we have in mathematics the feeling that we know there is something more.[16]

By virtue of this heterogeneity between the subject of experience and mathematical being we can affirm, drawing support from the strong autonomy thesis, that there is no necessary relation between the mathematical and the empirical determinations of being. True enough, examples can be produced to show a coincidence but these examples do not show that their probability is not null. Furthermore, they always allow us to suspect that there is 'something more'.

The most frequently given example is mathematical physics, but this example is precarious since it does not show the necessity of coincidence but rather it shows that mathematical knowledge has gone beyond its own domain to function philosophically as a sketch of the object of classical physics.

If nothing else, the above serves to prove that the essence of mathematics lies within the infinity toward which it points. It is in this 'pointing' that the transcendental character of mathematics is to be found.

There is no guarantee of coincidence. If any, it is, as Pascal pointed out, accidental: if an experience coincides with mathematics it is due to the fact that such an experience was already mathematical.

II.3

My last point has to do with the relation between mathematics and reality. To consider it, I must recall the difference between transcendent and transcendental:

1. The transcendent principles are defined in opposition to the immanent ones as those "which profess to pass beyond these limits [the limits of possible experience – S.R.]";[17] moreover, "there can never be any adequate empirical employment of the principle".[18] The transcendent is concerned with "an object of experience, but only in that aspect in which it ceases to be an object of experience".[19] When something "is posited in that which lies entirely outside the sensible world, and therefore outside all possible experience, the ideas become transcendent ... they detach themselves completely from experience, and make for themselves objects for which experience supplies no material ... ".[20]

2. Transcendental refers to the

> employment or misemployment of categories, which is merely an error of the faculty of judgement when it is not duly curbed by criticism, and therefore does not pay sufficient attention to the bounds of the territory within which alone free play is allowed to pure understanding. I mean actual principles which incite us to tear down all those boundary-fences and to seize possession of an entirely new domain which recognizes no limits of demarcation.[21]

Transcendental assertions consequently, demand for themselves an "insight into what is beyond the field of all possible experiences".[22] That is to say, into the transcendent.

Nonetheless, transcendental ideas "have their own good, proper, and therefore immanent use ... ".[23] It is not the idea, then, but its use which can or cannot be transcendental or immanent: human reason has a natural tendency to transgress its limits.[24]

Transcendental logic is built upon hope[25] and has the intention of directing understanding to a certain aim, to a *focus imaginarius*:[26]

> who could be satisfied with the new experimental knowledge in which all cosmological problems concerning the duration and the magnitude of the world of liberty or of natural necessity, if, no matter how we proceed, every answer given gives birth to a new question demanding a new answer showing, thus, the insufficiency, for the tranquility of reason, of all the physical modes of explanation? (...) who cannot see, in natural contingency and in the interdependence of everything that can be thought and accepted according to the principles of experience, the impossibility of staying within them and

does not feel the urge, despite all prohibitions, to get lost in transcendental ideas?[27]

NOTES

* A first version of this paper was read at the Boston Colloquium for the History and Philosophy of Science in October 1984.

1 Jean Cavaillès, 'L'école de Vienne au congrès de Prague', in *Rev. de Méta. et de Morale*, 1935.

2 'Le deuxième courses universitaires de Davos', in dossier Cavaillès, ENS.

3 Cavaillès, *Sur la logique et la théorie de la science*, p. 10.

4 Cavaillès, *Gouhier manuscript*. All other unspecified quotes are taken from here.

5 Does not exist.

6 *Sur la logique et la théorie de la science*, p. 13.

7 Following Lorraine Daston's commentary at the Boston Colloquium.

8 S. Ramírez, 'Jean Cavaillès and the Vienna Circle' in *Grazer Philosophische Annalen*, Vol. 27, 1986.

9 Jean Cavaillès was among the listeners. He wrote about the occasion: "Ce sont en effet des affirmations préliminaires pour celle-ci, que la nécessité de fonder la possibilité de toute connaissance ontique sur celle de la connaissance ontologique, que l'attribution à la pensée humaine d'un caractère essentiel de finitude – révété par l'angoisse – dont la connaissance permet seule de poser correctement les problèmes de l'Etre et du Néant, enfin que la définition de la vraie métaphysique". 'Le deuxième courses universitaires de Davos', in dossier Cavaillès, ENS.

10 E. Kant, *Critique of Pure Reason*, Bxiii.

11 Kant, *Critique of Pure Reason*, B741–756.

12 Delete.

13 M. Heidegger, *Kant and the Problem of Metaphysics*, 5.

14 Heidegger, *op. cit.*, 4.

15 Kant, *Critique of Pure Reason*, B741–742.

16 L. Wittgenstein, *Philosophical Investigations*, 209.

17 Kant, *op. cit.*, B352.

18 Kant, *op. cit.*, B365.

19 Kant, *op. cit.*, B427.

20 Kant, *op. cit.*, B593.

21 Kant, *op. cit.*, B352.

22 Kant, *op. cit.*, B453.

23 Kant, *op. cit.*, B671.

24 Kant, *op. cit.*, B670.

25 Kant, *op. cit.*, B51.

26 Kant, *op. cit.*, B672.

27 Kant, *Prolegomena*, 57.

ALEJANDRO R. GARCIADIEGO

THE PRINCIPLES OF MATHEMATICS OF BERTRAND RUSSELL*

A biobibliographical note on its composition. An attempt to bring together the bibliographical data relevant to the composition of The Principles of Mathematics *(May, 1903)*

I

Bertrand Russell wrote prolifically for a long period of time. He lived for almost ninety-eight years and wrote more than seventy books, hundreds of articles and thousands of letters. This material included his personal and intellectual biographies. In one of Russell's personal autobiographies he outlined, briefly and clearly, how he came to write *The Principles of Mathematics*:

> with the beginning of October [1900] I sat down to write *The Principles of Mathematics*, at which I had already made a number of unsuccessful attempts. Parts III, IV, V, and VI of the book as published were written that autumn. I wrote also parts I, II and VII at that time, but had to rewrite them later, so that the book was not finished in its final form until May 1902 (Russell 1967, p. 144).

Another very similar description is contained in his intellectual biography.[1] These two sources, suggest that Russell began to compose the 'first final draft' of *The Principles of Mathematics* in October 1900 and finished it, as he claimed, on the last day of the nineteenth century. I might add that there are surviving records from the time that *The Principles of Mathematics* was written that are consistent with Russell's statements. (These records are kept at The Bertrand Russell Archives [McMaster University, Ontario, Canada].) Moreover, in the printed preface of the book, originally written in December 1902 when his memories were still fresh, Russell outlined the intellectual path that he followed. He gave special emphasis to the philosophical and mathematical influences that affected him as he composed the book. These recollections seemed to be

S. Ramirez and R. S. Cohen (eds.), Mexican Studies in the History and Philosophy of Science, 213–234.

consistent with the citation above. There is plenty of evidence, not just autobiographical, scattered through both Russell's numerous writings and some secondary sources, that support Russell's recollection in one way or another.

Russell claimed that the central parts of the book (i.e., parts III to VI) were published almost 'exactly as [he] wrote them during the autumn of 1900'. Some of the folios belonging to these parts were dated, in Russell's handwriting, between November and December 1900, providing even stronger evidence in favor of Russell's recollections. So far, so good!

The reader might wonder why a discussion of the events surrounding the writing of *The Principles of Mathematics* is necessary, given that Russell's own description of these events is available, and, at first glance, this description seems to be accurate.

Nevertheless, Allan Wood, one of Russell's most popular and well-known biographers, claimed that he could find two opposing quotations from Russell concerning almost all his endeavors (Wood 1957, p. 212). Later, I will show that this is the case by presenting another description of the events surrounding Russell's composition of his famous treatise. Furthermore, Wood provided a description of the methodology used by Russell while preparing and writing a book. Wood's account, which is accurate, contradicted Russell's own account of the writing of *The Principles of Mathematics*. According to Wood,

> there were successive periods of intense thinking, each of which culminated in a book, which in the end, was written rapidly. Russell hardly ever revised anything he had written and almost never read a book after it had been published (Wood 1959, p. 203).

On the one hand, Wood raises the question of whether Russell's account of the events leading to the writing of *The Principles of Mathematics* quoted from his personal autobiography is accurate. Did Russell give a contradictory account elsewhere? On the other hand, according to Wood's quotation, he thought that Russell rarely revised the manuscripts. It would be expected, therefore, that the final version of *The Principles of Mathematics* would be almost identical to the earlier drafts. If so, these earlier manuscripts would have limited interest to historians intrigued by the development and metamorphosis of ideas. Secondly, Wood's observation seems to contradict Russell's own account of the preparation of his book: he claimed to have made several earlier attempts to write *The Principles of Mathematics*. Wood, however, assures us that Russell rarely made revisions. Therefore, several questions may intrigue professional historians. Among other things, these questions focus on the evolution of Russell's ideas from 1898 to 1903, as revealed by information contained in earlier attempts to write *The Principles of Mathematics*. Why were the initial attempts not successful? In this case, why did Russell make several attempts to write *The Principles of Mathematics*, even though, according to Wood, Russell rarely revised the work? Finally, was Russell's autobiographical account of his production of *The Principles of Mathematics* accurate, or did he give another description of these events elsewhere? Throughout this essay, I would like to emphasize the last two questions, which are highly relevant to a new historical interpretation of the events surrounding the writing of *The Principles of Mathematics*. I will base this alternative account on Russell's own contradictory statements on this issue concerning his own description. This new interpretation might help clarify the role played by the nascent set-theoretic paradoxes and, in addition, illustrate Russell's development of his *logicist thesis*. The set-theoretic paradoxes were discovered while Russell was writing *The Principles of Mathematics* and, the *logicist thesis* was the modular dissertation of the treatise.

Let me stress once more that my task is to present a plausible new reconstruction of the events surrounding the composition of the book. First it is necessary to elucidate the facts; only then can a comprehensive analysis of the evolution of ideas be offered.

Let me draw a picture of the final printed version of *The Principles of Mathematics*. This treatise was published in May 1903. It was finally structured, according to Russell's recollections, in the following way:

1. Preface (December 1902);
2. Part I: The indefinables of mathematics (May 1902);
3. Part II: Number (May 1902);
4. Part III: Quantity (October 1900);
5. Part IV: Order (November 1900);
6. Part V: Infinity & Continuity (November–December 1900);
7. Part VI: Space (December 1900);
8. Part VII: Matter & Motion (May 1902);
9. Appendix A: The logical & arithmetical doctrines of Frege (November 1902);

10. Appendix B: The theory of types (November 1902).

The last two sections were late additions to the main text, and are excluded from the present discussion.

II

A simple question arises from Russell's passage concerning the events surrounding the composition of *The Principles of Mathematics*: why did he have to rewrite only parts I, II and VII, but not the whole book? According to recent research, Russell's discovery of the set-theoretic contradictions (in particular, the ones labeled with his name and with Cantor's) forced him to reconsider his original ideas. The contradictions were closely associated with the most primitive concepts and relations of the theory of sets. This might explain the need to rewrite part I (on the indefinables) and possibly part II (on number). Nevertheless, it does not account for the reasons why he felt compelled to elaborate part VII (that on matter and motion, in which he discussed the foundations of a physical science: dynamics). On the contrary, it is logical to suppose that those subjects more closely linked with the set-theoretic contradictions needed further scrutiny, such as the concepts of infinity and continuity.

There are other primary sources that portray a slightly different picture of the events surrounding the preparation of *The Principles of Mathematics*. For example, in a discussion with Philip Jourdain, Russell stated that "during the rest of [1900] I wrote parts III to VI of my *Principles*". Then, he added in parentheses: "(Part VII is largely earlier, Part I and II wholly later, May 1902)" (Grattan-Guinness 1977, p. 133). It was already known that parts I and II had been written 'wholly later'; later on I will discuss how much later than 1900 these parts were written. At this time, however, Russell raises an additional issue. What does he mean, when discussing the composition of part VII, that this one was written *largely earlier*? Did he develop this part of the book even before he sat down to compose his now famous work in October 1900? This seems to be the case. Furthermore, there are personal letters written at the time that support this conjecture. In fact, part VII belonged to a previous version of *The Principles of Mathematics* written between July 1899 and June 1900, just few days before Russell left Cambridge for the First International Congress of Philosophy. If this is the case and if Russell had originally planned his book to contain seven parts, then he *did not* conclude his book on the last day of the last century: on that particular day, part VII was still a fragment of another manuscript. This reasoning is supported by several personal missives of Russell.

Nevertheless, it is also possible that Russell originally planned to have only six sections in his treatise. In this case, he finished writing the original six parts, as he claimed, on 31st December 1900. Perhaps he added the seventh part as an afterthought, just as he added appendices on Frege's work and on his solution to the set-theoretic paradoxes. But, even if this was the case, I would argue that Russell did not finish the original six sections by the above mentioned date.

III

It was at the First International Congress of Philosophy held in Paris in July of 1900, only a month after finishing his third and last attempt to write a book on the foundations of mathematics,[2] that Russell met Giuseppe Peano (1858–1932). Peano caused a turning point in Russell's intellectual life. Russell was impressed originally not so much with the article that Peano delivered at the Congress, but with Peano's mode of reasoning (in particular, his logic and symbolic language) which allowed him, according to Russell, to win all the debates on which he embarked. As a result of meeting Peano, Russell developed a different approach toward the foundations of mathematics.

Russell returned to England in mid-August of 1900 completely convinced that the work of the Italian mathematicians was superior to that of all their colleagues in matters relating to the philosophy of mathematics. Russell spent the months of September and October reading Peano's articles and learning the techniques that will allow him to express his thoughts without ambiguity. Russell studied other authors contained mainly in Peano's *Rivista di Matematica*. He also wrote an article for the same journal in which he extended the theory of relations to the notion of series (ordered successions); furthermore, he also reread and revised his earlier draft entitled *Principles of Mathematics*, before starting a new one.[3]

According to Russell's own description of the events surrounding the writing of *The Principles of Mathematics* – the one he narrates in his autobiography and is quoted at the beginning of the present essay – he sat down to write *The Principles of Mathematics* in October 1900, finishing it on December 31, 1900. Nevertheless, in a letter written on

September 11, 1917, Russell provides the following description of the writing of *The Principles*:

> I first came across Peano in 1900 at the Paris Congress of Philosophy, and it was he who started me on mathematical logic. It was only after mastering his work that I discovered how to treat arithmetic by mathematical logic. I wrote parts III to VI of *The Principles* in the autumn of 1900 (Grattan-Guinness, 1977a, p. 144).

This letter was written in 1917, at least 20 years *before* Russell wrote the first draft of the final version of his autobiography. And his description seems to be quite clear: he wrote parts III to VI that autumn, and there is no mention at all of parts I, II and VII.

This quotation, along with the quotation addressed to Jourdain (cited before), appears to contradict Russell's statement that he finished *The Principles of Mathematics* on December 31, 1900. However, the quotes perhaps should not be taken entirely literally. It needs to be emphasized that problems can arise in interpretation of Russell's autobiographical material due to his oversimplification of complex events. Russell was aware that his recollections would be read by academic specialists and the general public. Unfortunately, in order to simplify the description of certain ideas or events, some degree of distortion occurred, perhaps unwillingly introduced. For example, a detailed description of the writing of *The Principles of Mathematics* would have taken many paragraphs, or even a whole essay. It was easier to oversimplify this description and claim that the book was written in a specific period of time. Perhaps, too, by the mid-twentieth century, when Russell began to write his autobiography, his memory of these events had began to fade.[4]

Just as some historians have literally interpreted Russell's quotation suggesting that he finished *The Principles of Mathematics* on December 31, 1900, it is possible to interpret literally the quotation from the correspondence with Jourdain. I will discuss the second set of quotations, suggesting that Russell *only* wrote parts III to VI during the autumn of 1900. So, contrary to Russell's letter to Helen Flexner of December 31, 1900, I believe that Russell did not finish the first manuscript of *The Principles of Mathematics* on the last day of the last century.[5] How, though, could I deny the veracity of a primary source written at that time? There is, however, a second letter to Helen Flexner in which Russell informed her that he was working on the second volume of *The*

Principles of Mathematics and that "the work [he had] to do [was] nearly finished" (Letter: Russell to Helen Flexner, August 2, 1902). This letter was written on August 2, 1902, almost two years after his first letter. Nevertheless, the records kept at The Bernard Russell Archives show that the words *nearly finished* do not accurately reflect the state of the manuscript at the time.

Furthermore, I support the conjecture that parts I and II were *originally* written in May and June 1901. Although the table of contents of part I does not contain any reference to "The Contradiction", nonetheless the contradiction of the class of all classes which are not members of themselves is described in the second chapter of part I (see Table 1). This is contrary to the claim that the first discussion of the contradiction was written in May 1902 (cf. Blackwell, 1971a, p. 7; and Grattan-Guinness, 1978a, p. 135, note 4). Moreover, if Russell originally planned to include part VII, then this new interpretation still holds. Even if Russell wrote parts I and II in 1900, there is plenty of evidence that he *did not* write part VII that autumn. Therefore, the manuscript was not complete at the time.

In the section in which Russell discussed the irrational numbers, there is some evidence suggesting that Russell did not write parts I and II before November 1900. Russell says: "Arithmetical theories of irrational could be treated in parts I *or* II, since they depend essentially upon the notion of order" (Russell, 1900a, Part V, Chapter III, folio 40 (my emphasis)). In the printed version of the book Russell specified part II. The disjunction in writing this specific sentence in the manuscript might indicate that by November 1900, Russell had not decided whether to include his discussion of the theories of irrational numbers, which he had not yet written. It should be noted that even if Russell did not finish the whole manuscript of *The Principles* by December 31, 1900, he had finished parts III to VI, which contained almost five hundred folios, distributed in thirty-four chapters (an average of almost ten folios a day). This was a tremendous achievement in itself.

In the autumn of 1900, more precisely in November and December, Russell sat down to write yet another version of his 'big book '. Except for his autobiographical sources, there is no internal or external evidence indicating that Russell wrote parts I and II in those days. Although there are usually original folios of Russell's major works on record at The Bertrand Russell Archives, none survive from the draft of parts I and II

TABLE 1. This is an outline of the Table of Contents of the surviving folios of the first two parts of *The Principles of Mathematics*. Both parts were written in May and June 1901. Part II was published (in May 1903) as it was originally formulated. Part I had to be rewritten once Russell became completely convinced of the importance of his contradiction.

PART I	THE VARIABLE
Chapter I	The definition of pure mathematics.
Chapter II	Terms and concepts.
Chapter III	Classes and relations.
Chapter IV	Conjunction and disjunction.
Chapter V	The variable.
Chapter VI	Implication.
Chapter VII	Peano's symbolic logic.
Chapter VIII	Summary of Part I.
PART II	NUMBER
Chapter I	Definition of cardinal numbers.
Chapter II	Addition and multiplication.
Chapter III	Finite and infinite.
Chapter IV	Theory of finite numbers.
Chapter V	Addition of terms and addition of classes.
Chapter VI	Whole and part.
Chapter VII	Infinite wholes.
Chapter VIII	Ratios and fractions.

of *The Principles of Mathematics* that Russell claimed to have written in 1900.

It is been pointed out that there are other sources suggesting that Russell started by writing part III. Now, curiously enough, part III is missing from an even earlier draft finished approximately in June 1900, before Russell met Peano. At this point, according to Russell, the separation between the categories of number and quantity was completed; the category of quantity had lost its importance because all its concepts and theorems could be defined and deduced from the relation of order. He was simply discussing the topic for traditional reasons.

Some of the folios of parts III, IV, V and VI were dated, in Russell's handwriting, between November and December 1900. Those on part III were, perhaps, dated while Russell was revising this section. The last three parts of the book strongly reflected the influence of Cantor (see Table 2). There is additional evidence supporting the hypothesis that Russell started by writing part III. Most phrases that stress explicitly the

TABLE 2. The following Table of Contents summarizes the material that I believe Russell wrote in the last two months of 1900 (a folio of part VI is dated November 24, 1900). The reader should recall that I also claim that part VII was not composed at that time, but taken from an earlier manuscript (with some revisions and additions). The first folio of part III is dated November, 1900, but this seems to be a note added to indicate when Russell revised this section. All the corresponding folios are contained in the final draft of *The Principles of Mathematics*.

PART III	QUANTITY
Chapter I	The meaning of magnitude.
Chapter II	The range of magnitude.
Chapter III	Numbers as expressing magnitudes: measurement.
Chapter IV	Zero.
Chapter V	Infinity, the infinitesimal and continuity.
PART IV	ORDER
Chapter I	The genesis of series.
Chapter II	The meaning of order.
Chapter III	Asymmetrical relations.
Chapter IV	Difference of sense and difference of sign.
Chapter V	On the difference between open and closed series.
Chapter VI	Progressions and ordinal numbers.
Chapter VII	Dedekind's theory of numbers.
Chapter VIII	Distance.
PART V	INFINITY AND CONTINUITY
Chapter I	The correlation of series.
Chapter II	Real numbers.
Chapter III	Limits and irrational numbers.
Chapter IV	Cantor's first definition of continuity.
Chapter V	Ordinal continuity.
Chapter VI	Transfinite cardinals.
Chapter VII	Transfinite ordinals.
Chapter VIII	The infinitesimal calculus.
Chapter IX	The infinitesimal and the improper infinite.
Chapter X	Philosophical arguments concerning the infinite.
Chapter XI	The philosophy of the continuum.
Chapter XII	The philosophy of the infinite.
PART VI	SPACE
Chapter I	Dimensions and complex numbers.
Chapter II	Projective geometry.
Chapter III	Descriptive geometry.
Chapter IV	Metrical geometry.
Chapter V	Relation of metrical to projective and descriptive geometry.
Chapter VI	Dimensions of various spaces.
Chapter VII	The continuity of space.
Chapter VIII	Logical arguments against points.
Chapter IX	Kant's theory of space.

logicist thesis seem to have been added after the first versio of the book was finished (i.e., after parts III, IV, V and VI were written). In other words, I claim that Russell did not have as strong a commitment to the logicist thesis when he wrote the draft in autumn, 1900, as he did when he completed the 'final' version of *The Principles of Mathematics*. A reason that might explain his different attitude might be that fact that Russell may have changed his views after reading Cantor's seminal articles of 1895 and 1897 in November, 1900. As a result of reading these essays, in which Cantor constructed the finite cardinal numbers upon the concept of set, in addition to using techniques that he learned from Peano, Russell was finally able to demonstrate that the finite cardinal numbers were based on more primitive logical concepts.

IV

At this stage, when he started his new attempt to write in November 1900, Russell was convinced that all the apparent difficulties (*antinomies* is the word he used at this time) could be reduced to a simple problem of infinite number which could be resolved by a correct philosophy of 'any' (the existential quantifier). Russell now argued that the contradictions resulted from the supposition that well-defined infinite series have a last term. Russell now distinguished between infinite and finite numbers by arguing that the infinite numbers did not obey the principle of mathematical induction. Indeed, this is true since,

$$\aleph_n + 1 = \aleph_n \quad \text{(for any } \aleph\text{)}.$$

Nevertheless, Russell found a new difficulty a month later. He was convinced that there was a greatest transfinite cardinal number: that of the class of all individuals:

> But Cantor has given two proofs [. . .] that there is no greatest number. If these proofs be valid, there would seem to be still a contradiction. But perhaps we shall find that his proofs only apply to numbers of classes not containing all individuals [. . .] It is essential however, to examine this point with care, before we can pronounce infinity to be free from contradictions (Russell, 1900a, Part V, Chapter XII, folio 189).

However, at the time Russell wrote this part of *The Principles of Mathematics* in December 1900, he still thought he had solved all possible problems concerning the infinite, and concluded this section by asserting that "no definite contradictions can be proved concerning [the finite and continuity]" (*Ibid.*, folio 199). However, even before Russell had finished this manuscript, he had communicated to Louis Couturat the mistakes and fallacies found in Cantor. On December 8, 1900, Russell explained to his friend: "I have discovered a *mistake* in Cantor, who maintains that there is no largest cardinal number. But the number of classes is the largest number ..." (Letter: Russell to Louis Couturat, December 8, 1900; quoted and translated in Moore and Garciadiego 1981a, p. 325. Also quoted and translated in Coffa 1979a, p. 33).

Due to his attempt to clarify this mistake in Cantor, Russell came, eventually, to discover the set-theoretic paradoxes (Garciadiego 1983a, pp. 151–157; Garciadiego (1992, pp. 100–130)). In January 1901, Russell wrote his essay 'Recent works on the principles of mathematics' for the *International Monthly*. This essay was not well received by his former philosophy professors because Russell criticized their ignorance of the recent developments in the philosophy of mathematics. He also put forward some of the ideas that would later find their way into *The Principles of Mathematics*; in particular, he publicly claimed to have found a fallacy in Cantor's reasoning. This argument would be transformed later into one of the versions of the now famous Cantor's paradox. At this stage, however, Russell did not claim to have found a mathematical contradiction in Cantor's work. Russell was hoping to find a consistent explanation of the principles of mathematics and Cantor's theory played a major role in achieving Russell's goal. At the time, Russell thought that he was going to be able to show that Cantor was mistaken, but had not yet found any errors in Cantor's second proof, that there is no greatest cardinal number. Russell also proposed his logicist thesis, probably for the first time.

It is difficult to determine exactly what Russell was doing during the months of February, March and April of 1901, a few months before he discovered the 'contradictions'. Nevertheless, there is a folio containing the table of contents of an early version of part I of *The Principles of Mathematics*, in which 'the indefinables of mathematics' was discussed. It was probably written in April 1901 (see Table 3). This table of contents is somewhat similar to another table that Russell prepared approximately a month later. This second table of contents also outlines part II, dealing with 'numbers'. Nothing is mentioned, in any of the tables, of the possible existence of any contradictions (see Table 4).

TABLE 3. Russell drafted this Table of Contents of part I of *The Principles of Mathematics* in about April 1901. This Table of Contents is now kept in File 230-030350.

PART I	THE INDEFINABLES OF MATHEMATICS
Chapter I	Definition of pure mathematics.
Chapter II	Terms & concepts.
Chapter III	Propositions & assertions.
Chapter IV	Implication.
Chapter V	Propositions of given form.
Chapter VI	The variable.
Chapter VII	Conjunction, disjunction & negation.
Chapter VIII	Classes: extension & intension.
Chapter IX	Relations.
Chapter X	Definition.
Chapter XI	Peano's symbolic logic.

The first version of part I was written in May 1901. Although Russell segmented this first draft when he wrote a second version, Blackwell has been able to reconstruct it (see, Blackwell, 1971a; and Blackwell 1984–1985a). There are several important issues contained in this draft of 1901. First of all, the logicist thesis is completely described. But the most remarkable aspect of this draft of part I is that *it does contain a contradiction*. According to Russell,

> [w]e saw that some predicates can be predicated of themselves. Consider now those of which this is not the case. There are the referents (& also the relata) in what seems like a complex relation, namely the combination of non-predictability with identicity. But there is no predicate which attaches to all of them & to no other terms. For this predicate will either be predicable or not predicable of itself. If it is predicable of itself, it is one of those referents by relation to which it was defined, & therefore, in virtue of their definition, it is not predicable of itself. Conversely, if it is not predicable of itself, then again it is one of the said referents, of all of which (by hypothesis) it is predicable, & therefore again it is predicable of itself. This is a contradiction, which shows that all the referents considered have no exclusive common predicate, & therefore, if defining predicates are essential to classes, do

TABLE 4. In May 1901, Russell developed a tentative Table of Contents of the entire manuscript (kept in File 230-030350). The headings of parts III and IV are missing and the title of part VII is mentioned but the topic was not developed. Perhaps this is an indication that Russell ignored, at this time, the chapters composing part VII and that he did not write it (under the new influence of Peano) in December of 1900. Note that nothing has been written about the contradictions.

PART I	THE VARIABLE
Chapter I	The definition of pure mathematics.
Chapter II	Terms and concepts.
Chapter III	Classes and relations.
Chapter IV	Conjunction and disjunction.
Chapter V	The variable.
Chapter VI	Implication.
Chapter VII	Peano's symbolic logic.
Chapter VIII	Summary of Part I.
PART II	NUMBER
Chapter I	Definition of cardinal numbers.
Chapter II	Addition and multiplication.
Chapter III	Finite and infinite.
Chapter IV	Theory of finite numbers.
Chapter V	Addition of terms and addition of classes.
Chapter VI	Whole and part.
Chapter VII	Infinite wholes.
Chapter VIII	Ratios and fractions.
PART III	(QUANTITY)
Chapter I	Meaning of magnitude.
Chapter II	Range of magnitude.
Chapter III	Measurement.
Chapter IV	Zero.
Chapter V	Infinity, infinitesimal and continuity.
PART IV	(ORDER)
Chapter I	Genesis of series.
Chapter II	Meaning of order.
Chapter III	Asymmetrical relations.
Chapter IV	Difference of sense and of sign.
Chapter V	Open and close series.
Chapter VI	Progressions and ordinal numbers.
Chapter VII	Dedekind's theory of numbers.
Chapter VIII	Distance.

TABLE 4. (continued).

PART V	INFINITY AND CONTINUITY
Chapter I	Correlation of series.
Chapter II	Real numbers.
Chapter III	Limits and irrational numbers.
Chapter IV	Cantor's first definition of continuity.
Chapter V	Ordinal continuity.
Chapter VI	Transfinite cardinals.
Chapter VII	Transfinite ordinals.
Chapter VIII	The infinitesimal calculus.
Chapter IX	The infinitesimal and the improper infinite.
Chapter X	Philosophical arguments concerning the infinitesimal.
Chapter XI	The philosophy of the continuum.
Chapter XII	The philosophy of the infinite.
PART VI	SPACE
Chapter I	Dimensions and complex numbers.
Chapter II	Projective geometry.
Chapter III	Descriptive geometry.
Chapter IV	Metrical geometry.
Chapter V	Relation of metrical to projective and descriptive geometry.
Chapter VI	Dimensions of various spaces.
Chapter VII	The continuity of space.
Chapter VIII	Absolute and relative positions.
Chapter IX	Kant's theory of space.
PART VII	MATTER AND MOTION

> not form a class [...] It follows from the above, that not every definable collection of terms forms a class defined by a common predicate (Russell, 1901a, folios 22–23; see, also, Russell, 1902a, Part I, Chapter IX, folios 189–190).

At this point, this contradiction meant that there was a need for some kind of limitation when defining classes. Cantor had already arrived at the same conclusion a few years earlier (ca. 1872–1883, or 1896), probably as soon as he had proven his theorem claiming that the cardinality of the power set was always greater than the cardinality of the original set. Cantor also concluded that there was a need for some kind of limitation in defining infinite totalities; from here he probably introduced his distinction between inconsistent and consistent multiplicities. Russell's passage raises two remarkable points. First, and most important, Russell describes his contradiction of the class of all classes which are not members of themselves. Second, both Russell and Cantor had probably begun thinking about contradictions arising from the theorem proved by Cantor. Furthermore, as a result of this contradiction, Russell was able to deny one of Peano's most basic principles: "Peano holds [...] that every proposition containing only one variable is reducible to the form 'x is an a'. But we have already seen (Chapter II [*sic.*]) at least one such proposition is not reducible to this form" (Russell, 1901a, folio 56).

The earliest existing version of part II, that on 'Number', was written in June 1901. This part of *The Principles of Mathematics* clearly shows the impact of Cantor's influence on Russell. Through the text of this section, there are several footnotes referring to Cantor's work, especially his seminal papers of 1895 and 1897, in which Cantor finally expressed the principles of the theory of his transfinite numbers in a clear and concise (although perhaps also disappointing) way.

It was about this same time, June 1901, that Russell was intellectually involved with three mathematical contradictory arguments. First, whereas Cantor's theorem stated that there was no greatest class, Russell insisted that there was a class of classes, or universal class. (This argument would be transformed into Cantor's contradiction.) Second, there was also the contradiction derived from his attempt at reconciling Cantor's proof with his notion of the class of all classes, which was related to Peano's principle of the formation of classes, described in his May 1901 draft of part I. (This second argument would be transformed into Russell's contradiction.) Third, there were two theorems with apparently conflicting implications. One was demonstrated by Burali-Forti and the other by Cantor on the trichotomy law for transfinite ordinal numbers. (This third argument was associated with the seed of the future development of the Burali-Forti paradox.) Russell did not appear to have appreciated the importance of the implications of these arguments. He seems to have failed to grasp their importance at the time, nor did he communicate them to anyone. As time passed, it is reasonable to suppose that he felt very uncomfortable with these results. Although he was very close to finishing his book, he was now finding new problems without answers.

By June 1901, *The Principles of Mathematics* was almost complete, except for the part on 'Matter and Motion'. But Russell did not write it at this time. If, as other historians and philosophers claim, Russell did write Part VII in December 1900, then why did he have to rewrite it later? And if he did rewrite Part VII, why did he not rewrite parts IV, V

TABLE 5. Russell drafted this outline of part I of *The Principles of Mathematics* in April 1902 (now kept in File 230-030350). This section of the manuscript would be modified again, largely due to Russell's discussion with Frege.

BOOK I	THE VARIABLE
Chapter I	Definition of pure mathematics.
Chapter II	Symbolic logic.
Chapter III	Implication and formal implication.
Chapter IV	Proper names, adjectives and verbs.
Chapter V	Denoting.
Chapter VI	Classes.
Chapter VII	Assertions.
Chapter VIII	The variable.
Chapter IX	Relations.
Chapter X	The contradiction.

and VI, in which questions related to the contradictions of 'infinity' and 'continuity' were discussed?

By April 1902, perhaps after several unsuccessful attempts to resolve the difficulties, Russell decided to rewrite part I (see Table 5). This time he wrote a special discussion about his unresolvable contradiction. A few weeks later, during May 1902, he finished this section of his book devoting a whole chapter to a rigorous examination of the contradiction. In the final days of that month he had finished the book after he reread and revised a 'pile of old Ms' which contained a section on 'Matter and Motion'.[6]

V

So, by the end of May 1902, Russell finally completed his book on the principles of mathematics (see Table 6). But now he had encountered some irresoluble contradictions that he thought to be trivial at the beginning. In mid-June 1902, Russell communicated his discovery of the contradiction to Gottlob Frege (1848–1925). He had previously written to Peano, but received no answer. Russell continued to exchange letters with Frege and read his work. As a result of their discussions, Russell decided to include an appendix, explaining Frege's principles, and explaining how Frege's views differed from his own. There was yet another appendix: this one contained a potential, although unsuccessful, solution to the difficulties contained in the book. Both appendices were

TABLE 6. This table describes the table of contents and the present condition of the final draft of *The Principles of Mathematics* (May 1903), mainly finished in December 1902; although some folios were added in January 1903, and there were even some changes introduced presumably in galley. The manuscript is incomplete, there are several folios missing (indicated by the symbol *) from parts IV and VI. The file number is 230.030350, and it is divided in 20 folders. There were folios added from previous attempts: the old numeration is indicated in parenthesis.

	Contents	Folios	New & old numeration	1898	1899	1899–90	1901	Folder
	PREFACE	1–6						F1
PART I	THE INDEFINABLES OF MATHEMATICS							
Chapter I	Definition of pure mathematics	1–15						
" II	Symbolic logic	16–32a–60a–64	53(58)–62(68)				"	
" III	Implication and formal implication	65–81						F2
" IV	Proper names, adjectives and verbs	82–108	85(9),89(13)–91(15)				"	
" V	Denoting	103–126	105(16),106(17),111(26)–119(34)				"	
" VI	Classes	127–157	135(44),136(45),137(48),138(49)				"	F3
" VII	Propositional functions	158–169						
" VIII	The variable	170–184	174(37)				"	
" IX	Relations	185–194	185(18)–190(23)				"	
" X	The contradiction	197–204,1–3,204a–211						F4
PART II	NUMBER							
Chapter XI	Definition of cardinal numbers	1–12						
" XII	Addition and multiplication	13–19						
" XIII	Finite and infinite	20–25						F5
" XIV	Theory of finite numbers	26–34	49(3)			"		
" XV	Addition of terms and addition of classes	35–53						
" XVI	Whole and part	54–64	60(5)–61(6)			"		
" XVII	Infinite wholes	65–76	72(30),73(31),75(38)			"		
" XVIII	Rations and fractions	77–86						F6
PART III	QUANTITY							
Chapter XIX	The meaning of quantity	1–28						
" XX	The range of quantity	29–33a–37	29(11)–37(19)		"			
" XXI	Numbers as expressing magnitude: measurement	38–45a–54	45a(27),49(30),50(31),53(22)		"			F7
" XXII	Zero	55–63						
" XXIII	Infinity, the infinitesimal, and continuity	64,64a,64b–67a–80	67(34)–69(36),74(38)–75(39)		"			F8

TABLE 6. (Continued).

	Contents	Folios	New & old numeration	1898	1899	1899–90	1901	Folder
PART IV	ORDER							
Chapter XXIV	The genesis of order	1–17						F9
" XXV	The meaning of order	18–42						
" XXVI	Asymmetrical relations	41–46a–52,•,54–61						
" XXVII	Difference of sense and difference of sign	62–65a–72						
" XXVIII	On the difference between open and closed series	•						
" XXIX	Progressions and ordinal numbers	•						
" XXX	Dedekind's theory of number	?,108–112						F10
" XXXI	Distance	113–122						
PART V	INFINITY AND CONTINUITY							
Chapter XXXII	The correlation of series	1–25a						
" XXXIII	Real numbers	26–35a						
" XXXIV	Limits and irrational numbers	36–39a–43a–58						F11
" XXXV	Cantor's first definition of continuity	59–78	76(29)–77(30)		"			
" XXXVI	Ordinal continuity	79–84a–93						
" XXXVII	Transfinite cardinals	94–101a–104a–105						F12
" XXXVIII	Transfinite ordinals	106–108a–118a–123	107(47)–108(48)		"			
" XXXIX	The infinitesimal calculus	124,124a–134						F13
" XL	The infinitesimal and the improper infinite	135–146						
" XLI	Philosophical arguments concerning the infinitesimal	147–157a,157b–158						
" XLII	The philosophy of the continuum	159–174						
" XLIII	The philosophy of the infinite	175–199						F14

TABLE 6. (Continued).

	Contents	Folios	New & old numeration	1898	1899	1899–90	1901	Folder
PART VI	SPACE							
Chapter XLIV	Dimensions and complex numbers	1–19						
" XLV	Projective [space] geometry	20–44						F15
" XLVI	Descriptive geometry	45–54a–66						
" XLVII	Metrical geometry	67–81a						
" XLVIII	Relation of metrical to protective and descriptive geometry	•						F16
" XLIX	Definitions of various spaces	•						
" L	The continuity of space	•						
" LI	Logical arguments against points	?–169–173						
" LII	Kant's theory of space	174–187	180(65),182(67)–185(70)					F17
PART VII	MATTER AND MOTION							
Chapter LIII	Matter	1–11						
" LIV	Motion	11a–16a–21						F18
" LV	Causality	22–34						
" LVI	Definition of a dynamical world	35–38						
" LVII	Newton's laws of motion	39–59	39(35)–48(44),50(47)–55(52),57(53)–58(54)					
" LVIII	Absolute and relative motion	60–68						F19
" LIX	Hertz's dynamics	69–76,1–3	69(54a)–76(61)					
APPENDIX A	THE LOGICAL AND ARITHMETICAL DOCTRINES OF FREGE	1–23a–34a–44						F20
APPENDIX B	THE DOCTRINE OF TYPES	1–9						

received by Cambridge University Press on November 15, 1902 (*Ibid.*, Part VII, Chapter III, folio 30). Less than a month later, the publisher received the preface of the book; and during January 1903, Cambridge University Press received the last folios for which an approximate date of composition is known. Other changes (for example, the addition of sections §299, §300 and §301) were made, but there is no external evidence on which to base an approximate data of composition of these last sections. The addition of these particular sections is important, because they contain the arguments from which the Burali-Forti paradox would later be derived (see Garciadiego, 1983a, pp. 182–184 and Moore and Garciadiego, 1981a, pp. 327–331).

To conclude, it is known that the main body of *The Principles of Mathematics* was finished by December 1902 – except for minor corrections. The book was finally published in May 1903 in spite of Russell's own terrible and deep disappointment (Letter: Russell to Helen Flexner, May 13, 1903; letter: Frank Russell to Bertrand Russell, May 9, 1903, The Bertrand Russell Archives; letter: Russell to Alys Russell, May 16, 1902; letter: Russell to L. Dickinson, August 2, 1902, printed in Russell, 1967a, p. 184; letter: Russell to Gilbert Murray, December 28, 1902, printed in Russell, 1967a, p. 163). Russell summarized his feelings towards the completion of the book in the following terms.

> My book is nearly out, when it appears I shall feel as one does when one leaves the dentist. I do not think the book good, but it will be done, and that is enough (Letter: Russell to Helen Flexner, February 7, 1903, The Bertrand Russell Archives).

NOTES

* This essay was written in 1984. Different previous versions of the paper, treating various issues relating to the composition of *The Principles of Mathematics*, were presented at the XIX National Congress of the Mexican Society of Mathematics (1983), the Annual Meeting of the American Mathematical Society (Phoenix, Arizona, 1989) and the XVIII International Congress on History of Science (Hamburg and Munich, 1989). An extended, updated and detailed discussion of all issues contained in this essay is presented in: Alejandro R. Garciadiego, *Bertrand Russell and the Origin of the Set Theoretic Paradoxes*. (Birkhäuser).

[1] "I spent October, November and December of [1900] on *The Principles of Mathematics*. Parts III, IV, V and VI of that book are almost exactly as I wrote them during those months. Parts I, II and VII, however, I re-wrote later. I finished this first draft of *The Principles of Mathematics* on the last day of the nineteenth century – i.e. 31st December, 1900" (Russell 1959a, p. 56).

[2] These manuscripts are entitled: *An analysis of mathematical reasoning, The fundamental ideas and axioms of mathematics* and *Principles of Mathematics.* For a discussion of their contents, see Garciadiego 1992.
[3] In October, 1900, Russell wrote the following comments on the first folio of part II (Whole & Part) of a manuscript simply entitled *Principles of Mathematics,* which he had finished the previous June. "I must preface Arithmetic, as Peano does, by the *true log[ical]* Calc[ulus] to be called *Book I,* the Individual". "Note, I have been wrong in regarding the Log. Calc. as having specially to do with whole & part. *Whole* depends on three notions: (1) implication, (2) &, and (3) negation. Whole & part requires the Teoria della Grandetze, i.e., a special form of addition, not that of Log. Calc."
[4] Although it is well known that Russell was extremely fit, both physically and mentally, even near the end of his life, in his recollections he confused the timing of major events (see Russell, 1956a, pp. 195–196; Blackwell, 1973a, p. 13).
[5] In his intellectual autobiography (Russell, 1959a), Russell claims to have finished a first draft by 1900. Some researchers were influenced by Russell's use of the word *first,* along with Russell's subsequent revisions; they suggested that Russell wrote both a penultimate and a final of *The Principles of Mathematics* (see Blackwell, 1971a; Grattan-Guinness, 1978a; and Griffin, 1986–1987a).
[6] You "will be surprised & amused, after all my talk of two months, to hear that I finished my book yesterday. I found that a pile of old Ms, which I had expected to have to re-write, required only a few additions and corrections, so I arrived at a sudden termination" (Letter: Russell to Alys (Russell's first wife), May 24, 1902. The Bertrand Russell Archives).

REFERENCES

Blackwell, K. 1971. *The text of Russell's The Principles of Mathematics.* The Bertrand Russell Archives. Typescript.

Blackwell, K. 1973. 'Our knowledge of *Our Knowledge*'. *Russell: The Journal of the Bertrand Russell Archives* **12**, 11–13.

Blackwell, K. 1984. 'Part I of *The Principles of Mathematics*', *Russell: The Journal of the Bertrand Russell Archives* N. S. **4**, 271–288.

Coffa, A. 1979. 'The humble origins of Russell's paradox'. *Russell: The Journal of the Bertrand Russell Archives* **33–34**, 31–38.

Garciadiego, A. 1981. See, Moore and Garciadiego, 1981.

Garciadiego, A. 1983. *Bertrand Russell and the Origin of the Set Theoretic Paradoxes,* Ph.D. Dissertation. University of Toronto.

Garciadiego, A. 1986. 'L'influence de Georg Cantor sur Bertrand Russell'. *Cahiers du Séminaire d'Histoire des Mathématiques de Toulouse* **8**, 1–14.

Garciadiego, A. 1992. Bertrand Russell and the origins of the set theoretic paradoxes. Basel: Birkhäuser.

Grattan-Guinness, I. 1977. *Dear Russell – Dear Jourdain.* London: Duckworth.

Grattan-Guinness, I. 1978. 'How Bertrand Russell discovered his paradox'. *Historia Mathematica* **5**, 127–137.

Griffin, N. 1986–1987. 'Victor Lowe, Vol. I of *Alfred North Whitehead. The Man and His Work*'. *Russell: The Journal of the Bertrand Russell Archives* N.S. **6**, 172–178.

Moore, G. and Garciadiego, A. 1981. 'Burali-Forti's paradox: A reappraisal of its origins'. *Historia Mathematica* **8**, 319–350.

Russell, B. 1898, *An Analysis of Mathematical Reasoning[,] being an Inquiry into the Subject-Matter, the Fundamental Conceptions, & The Necessary Postulates of Mathematics*. The Bertrand Russell Archives. Manuscript.

Russell, B. 1899. *The Fundamental Ideas & Axioms of Mathematics*. The Bertrand Russell Archives. Manuscript.

Russell, B. 1899–1900. *Principles of Mathematics*. The Bertrand Russell Archives. Manuscript.

Russell, B. 1900. *The Principles of Mathematics* [parts III to VI], contained in: Russell, 1902. The Bertrand Russell Archives. Manuscript.

Russell, B. 1901. *Part I: The Variable*. The Bertrand Russell Archives. Manuscript. (Reconstructed according to Blackwell, 1971a, p. 7).

Russell, B. 1902, *The Principles of Mathematics*. The Bertrand Russell Archives. Manuscript. ["Finished May 23, 1902"].

Russell, B. 1956, *Portraits from Memory*. London: George Allen & Unwin.

Russell, B. 1959, *My Philosophical Development*. London: Unwin Books (paperback edition).

Russell, B. 1967. *The Autobiography of Bertrand Russell, 1872–1914*. London: George Allen & Unwin.

Spadoni, C. 1977. *Russell's Rebellian against Neo-Hegelianism*. Ph.D. Dissertation. University of Waterloo.

Wood, A. 1957. *Bertrand Russell, The Passionate Sceptic*. London: Unwin Books (paperback edition).

Wood, A. 1959. 'Russell's philosophy: A study of its development', contained in Russell, 1959, pp. 189–205.

GUILLERMO ZAMBRANA CASTAÑEDA

WITTGENSTEIN ON MATHEMATICAL PROOF

If one is willing to speak about mathematical proof in Wittgenstein, one must take into account that for him, in philosophy, "... we are not laying foundations, but tidying up a room, in the process of which we have to touch everything a dozen times. The only way to do philosophy is to do everything twice".[1] Thus, it does not come as a surprise that several times he goes over the same subject, and that it is practically impossible to obtain, by this token, a unique solution, a definite idea, a great global vision of the questions he poses.

Regarding the discussion of mathematical proof, we might begin by analyzing the cluster of questions Wittgenstein proposes in relation to the possibility of writing, one after the other, in an orderly fashion, at a command, the numbers making up the series of even natural numbers.[2] This is to say: how do I know which number follows the preceding in this series? And if I know beforehand, how do I know what to do with this knowledge when I have, in fact, to write down the sign? What do I do with the certainty that I can continue?

The matter at hand will be linked by Wittgenstein to the discussion of the answer to the question: "But then what does the peculiar inexorability of mathematics consists of?" Namely: "Would not the inexorability with which two follows one and three follows two be a good example?"[3] This response, given as a new question, points, nonetheless, to inexorability as the theme through which basic concepts that allow a better understanding of Wittgenstein's position on mathematics, will arise.

What stands out, then, is an approach in which convention (given by the use we make of words and signs), and training, explain both what we call inexorable as well as what we call truth in mathematics. After the fact that one number follows another depending on the series we are working on, Wittgenstein then asks if this means that it is correct to count anyway one pleases, he says:

> We should presumably not call it 'counting' if everyone said the numbers one after the other *anyhow*; but of course it is not simply a question of a name. For what we call 'counting'

S. Ramirez and R. S. Cohen (eds.), Mexican Studies in the History and Philosophy of Science, 235–248.

> is an important part of our life's activities ... Counting (and that means: counting like *this*) is a technique that is employed daily in the most various operations of our lives. And that is why we learn to count as we do: with endless practice, with merciless exactitude; that is why it is inexorably insisted that we shall say 'two' after 'one', 'three' after 'two' and so on. – 'But is this counting only a *use*, then: isn't there also some truth corresponding to this sequence?' The *truth* is that counting has proved to pay. – 'Then do you want to say that 'being true' means: being usable (or useful)?' No, not that, but that it can't be said of the series of natural numbers – any more than of our language – that it is true, but: that it is usable, and, above all, *it is used*.[4]

In relation with this insistence on the conventional nature of mathematics, we must take into account Wittgenstein's position.

> Now can it be said that the concepts which mathematics produces are a convenience, that essentially we could do without them?
> First and foremost the adoption of this concept expresses the sure expectation of certain experiences. *We do not accept* e.g. a multiplication's not yielding the same result every time.
> And what we expect with certainty is essential to our whole life.[5]

Having said this, inexorability and truth will now be associated by Wittgenstein in a very specific manner to logical necessity, logical inference, and use.

> 'But doesn't it follow with logical necessity that you get two when you add one to one, and three when you add one to two? And isn't this inexorability the same as that of logical inference?' – Yes! It is the same – 'But isn't there a truth corresponding to logical inference? Isn't it true that this follows from that?' – The proposition: 'It is true that this follows from that' means simply: this follows from that. And how do we use this proposition? – What would happen if we made a different inference – how should we get into conflict with truth?[6]

Wittgenstein now proposes the consideration of two cases, one the handling and usage of a ruler made out of a very soft rubber, the other of a ruler capable of expanding to a extraordinary extent if slightly heated.

In both cases we might say that the rulers are unusable, but then again, we surely can find a use (a way to measure with these rulers) capable of fulfilling some practical purpose, and we could say that "... What is here called 'measuring' and 'length' and 'equal length', is something different from what we call those things. The use of these words is different from ours; but it is *akin* to it; and we too use these words in a variety of ways".[7]

In this case, the 'truth' of line measurement made with the flexible or rigid ruler, depends on the circumstances under which I measure, this is to say, there is a conflict with truth when I use a flexible ruler when I want or need a common measurement, valid for every day life, and *vice versa*. It is in this way that inference and inexorability are referred to the system in which we measure. It is inexorable, and true, that two things can be exactly joined if they measure exactly the same when measured with the same ruler; as it is inexorable, and true that they cannot if measured once with an elastic, stretched ruler and then with a rigid one. Now the sense is clear in which the proposition 'It is true that this follows from that' only means 'this follows from that'.

If one is willing to think quite radically, the last proposition acquires a strictly declarative character, in which the 'problem' of truth has been relegated to a second plane. Quite a number of examples can be found in mathematics: one has only to change from one theory to another for mathematical objects to have a different sense; for mutually contradictory results to be, each on its own, considered true. One has only to think of the consequences implied by the acceptance or not of Euclid's fifth postulate, or of the completely different algebraic properties of (a, b) as member of RXR and $a + bi$ as a member of Complex Numbers, even if they can be thought of as the same point in space.

We will now turn to a more detailed account of Wittgenstein's discussion of inference.

> We must get clear what inferring really consists in: We shall perhaps see it consists in the transition from one assertion to another. But does this mean that inferring is something that takes place when we are making a transition from one assertion to another, and so *before* the second one is uttered – or that inferring consists in making the one assertion follow upon the other, that is, e.g., in uttering it after the other? (...) let's look at what happens here. – There is a transition from one

> proposition to another via other propositions, that is, a chain of inferences (...) *There is nothing occult about this process: it is the derivation of one sentence from another according to a rule: a comparison of both with some paradigm or other, which represents the scheme of the transition: or something of the kind.* This may go on paper, orally, or 'in the head'. – The conclusion may however be drawn in such a way that the one proposition is uttered after the other, without any such process; or the process may consist merely in our saying 'Therefore' or 'It follows from this', or something of the kind. We call it a 'conclusion' when the inferred proposition *can* in fact be derived from the premise.[8]

We can now say that the main point, and one of the key elements for understanding what Wittgenstein says about mathematics, is the fact that once 'inference' has been unlinked from 'truth', then following a rule and comparison with a paradigm (and of course, the acquisition of both) will be fundamental components of the inferring process.

Furthermore, we can add to the last passage some considerations regarding Wittgenstein's conceptions of the theme that will give us clearer context in which it is possible to understand his complementary conceptions, propositions and suggestions concerning inference.

> What we call 'logical inference' is a transformation of our expression. For example, the translation of one measure into another. One edge of a ruler is marked in inches, the other in centimetres. I measure the table in inches and go over to centimetres *on the ruler*. – And of course there is such a thing as right and wrong in passing from one measure to the other; but what is the reality that 'right' accords with here? Presumably a *convention*, or a *use*, and perhaps our practical requirements.[9]

It is now particularly important to think about two aspects: what is implied by the fact that a conclusion can be reached through a rule; and that what follows can really be inferred from given rules. This raises two possibilities: inference is done through calculation (for which we have been trained).[10] Or thanks to the agreement of rules with a reality (a logic) that seems abstract, very general, and quite rigid.[11]

Thus we learn to infer.[12] It is in practice and use that we learn the meaning of 'all'; "... by learning that '$f(a)$' follows from '$(x) \cdot fx$'".[13]

Its importance arising from the fact that in our language 'all' is a fundamental concept,[14] one that does not admit an exception.[15]

In face of the questions: do we always infer with the same laws? Is inferring a special activity?[16] Wittgenstein insists

> . . . it is necessary to look and see how we carry out inferences in the practice of language, what kind of procedure in the language-game inferring is.
> For example: a regulation says 'All who are taller than five foot six are to join the . . . section'. A clerk reads out the men's names and heights. Another allots them to such-and-such sections. – 'N.N. five foot nine'. 'So N.N. to the . . . section'. That is inference.[17]

Let us go back, after these remarks, and look into the writing of the series of even numbers. We can explain in what sense the fact is to be understood that in a series (the $A+2$ series, for example) transitions are given beforehand, and we can explain why we know how to continue.

As long as there is a sure way for determining the transitions that someone must carry out before he actually does them, we can always come up with a method that guarantees this, that is, we can delineate the series very tenuously, we can dictate it, etc. And we can, also, train people, in the usual sense that training and teaching have, so they will do things in the same manner and with similar results (as in multiplication). Then, if in this way we have the certainty that the transition from one term to the other will be correct, we could use the *expression* that the steps have already been given, as a picture (*Bild*) that describes the fact.[18]

From another perspective, and in relation to the certainty with which we infer one proposition from another, in face of the positiveness with which we say that some propositions follow from others, Wittgenstein says: ". . . – We ascertain that what is written here follows from what is written there. And this proposition is being used *temporally*". This idea of inference as a temporal event will further play on important role.

The series of questions previously discussed by no means exhaust the matter at hand; it will be expanded, and referred to a key element of mathematics: mathematical proof.

> When I say 'this proposition follows from that one', that is to accept a rule. The acceptance is *based* on the proof. That is to say, I find this chain (this figure) acceptable as a *proof*. – 'But

> could I do otherwise? Don't I *have* to find it acceptable?' – Why do you say you have to? Because at the end of the proof to say e.g.: 'Yes – I have to accept this conclusion'. But this is after all only the expression of your unconditional acceptance. I.e. (I believe): the words 'I have to admit this' are used in two *kinds* of cases: when we have got a proof – and also with reference to the individual steps of the proof.[19]

If it is precisely around the individual steps of the proof that the previous discussion was conducted, we have yet to see where a discussion of proof will take us, we have yet to see how it acquires its meaning and sense. As we had anticipated, Wittgenstein will offer several approximations to the matter.

He will consider first that what is important is to analyze what one does with the proof. He says:

> But how about when I ascertain that the pattern of lines in Figure 1a is like numbered with the pattern of angles in Figure 1b. (I have made the patterns memorable on purpose) by correlating them in Figure 1c. Now what do I ascertain when I look at this figure? What I see is a star with threadlike appendages.[20]

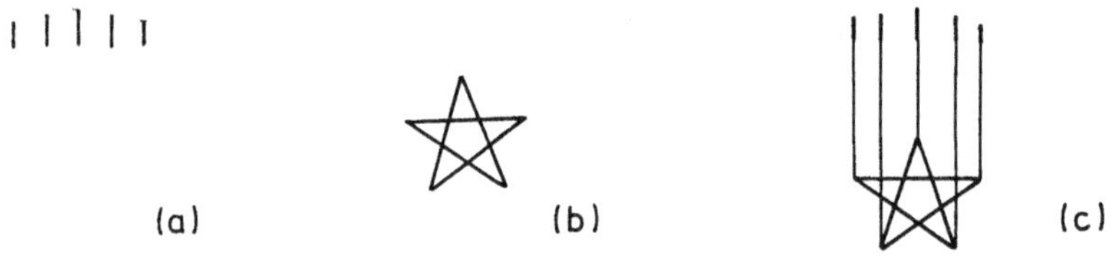

Fig. 1.

But then, if we think that the Figure 1(b) stands for five people arranged in a pentagon, Figure 1(a) as wands against the wall, Figure 1(c) would stand for the fact that I can give each of the people a wand. "I could regard Figure 1(c) as a schematic *picture* (*Bild*) of my giving the five men a wand each".[21]

For Wittgenstein, drawing the lines that correlate figures 1(a) and (b), thought of as a *temporal* act, cannot be considered a mathematical proof; nonetheless, Figure 1(c) proves that Figure 1(a) has as many wands as the star in 1(b) has points. This proposition, Figure 1(c), is *non-temporal*.[22]

Here, we can emphasize the distinction between what is temporal and what is non-temporal. The process of proving, the transitions (inferences) made from one proposition to the next, done temporally, yield a non-temporal 'result' whose logical validity is now fixed.

Thus, mathematical proof is thought of as a single pattern, with a beginning and an ending, of which the 'design' can be described (that is, the chain of inferences between beginning and ending); and which once obtained, shows us that it makes sense that a proposition *p* is, in fact, derived from *q* and *r*.[23]

A proof, then, will now be a paradigm that serves to certify or deduce what happens with different things under similar circumstances. In the example we are concerned with, Figure 1(c) allows us to say that if we have object sets arranged as (a), and (b), they are equal in number.[24] Furthermore, insofar this proof is used to compare any two groups as (a) and (b) without carrying out the actual correlation, it is the guarantee that (a) and (b) are equal in number.[25]

Once we have a proof, we act in one way and reject any other.[26]

On a second way of considering the mathematical proof, Wittgenstein proposes, quite radically:

> I might also say as a result of the proof: 'From now on an (a) and a (b) are called <<the same in number>>'.
> Or: the proof doesn't *explore* the essence of the two figures, but it does express what I am going to count as belonging to the essence of the figures from now on. – I deposit what belongs to the essence among the paradigms of language.
> The mathematician *creates essences.*[27]

But he immediately sets a boundary: "... it is not the *property* of an object that is ever 'essential', but rather the *mark of a concept.*"[28]

That is to say, this non-temporal product of a proof (in contrast with the temporal process of proving) is not an ontological description of the world; it is barely the mark of a concept, at the most an epistemological approach to this same world.

Giving examples of this thesis is fundamental for Wittgenstein. He proposes then to imagine one hundred marbles, numbered and arranged in ten groups, each group made up of ten marbles; each group is further arranged beneath another and with a small separation between five consecutive marbles. We must think, also, that the entire process of doing

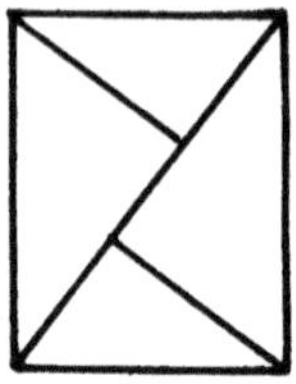

Fig. 2.

this with the marbles, which can be thought of as an experiment, has been filmed and that we see it on a screen; he adds:

> ... what I now see on the screen is surely not an experiment, for the picture of an experiment is not in itself an experiment. – But I see the 'mathematically essential' thing about the process in the projection too! For here there appear first a hundred spots, and then they are arranged in tens, and so on and so on.
> Thus I might say: the proof does not serve as an experiment; but it does serve as the picture of an experiment.[29]

By way of a mathematical example, we shall mention, briefly, that a sense very similar to the one offered by this perspective can be obtained from reading the way in which Georg Cantor defines 'aggregate' (*Menge*), its 'elements', union of such aggregates, 'partial aggregates', 'power' or 'cardinal number', etc., and goes on to construct a whole theory of Transfinite Numbers.[30]

Going back to Wittgenstein; from yet another standpoint, he raises new questions relating to mathematical proof.

When we think of a rectangle, divided in triangular pieces as shown in Figure 2, as the successful solution to a puzzle, we could say that we found a position for the triangles which we had not thought of beforehand: we discovered that they can be arranged in such a manner.[31] Even if, looking at the solution we say that we had never thought that these triangles could be laid out in that way.[32] Even more, Wittgenstein thinks it possible that there might be something in the puzzle's parts that somehow prevents us from arranging them in the position given by the solution: that we try other positions than the desired one; that, of course, the parts will not take care of themselves; that

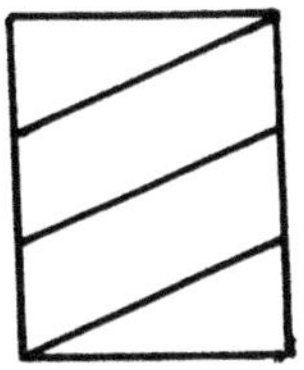

Fig. 3.

> ... This position is as it were excluded from space. As if it there were e.g. a 'blind spot' in our brain here. – And *isn't* it like this when I believe I have tried all *possible* arrangements and have always passed this one by, as if bewitched?
> Can't we say: The figure that shows you the solution removes a blindness, or again changes your geometry? It as it were shows you a new dimension of space. (As if a fly were shown the way out of the fly-bottle.)[33]

Working on the same kind of ideas, Wittgenstein considers the following proposition:

A rectangle can be made of two parallelograms and two triangles.
Proof: see Figure 3.[34]

Even if it seems as if we are compelled to accept this proposition as true, Wittgenstein argues for the possibility that a child might find surprising that the skewed sides of a parallelogram can form a straight line, and that the child might find it very difficult to produce a rectangle from the proposed figures, to the extent that this arrangement could be seen as the product of a magical trick, of something unnatural; as if the parallelograms had changed their nature.[35]

But we admit the figure as a proof, even if what it reveals can surprise us.[36] This is, that without thinking, we paste up the parallelogram on the top to the one beneath it and see the result.[37]

Then, "When one says: 'This shape consists of this shape' – one is thinking of the shape as a fine drawing, a fine frame ... on which, as it were, things which have this shape are stretched";[38] it is in this manner that for Wittgenstein, it is shown "... the essential property of this shape.

– You have shown me a new *picture* (*Bild*) ... 'The proof has taught me – shown me – that this is the case"'.[39]

This reference to essentiality must be seen in the perspective pointed out above; what is essential, in this case also, is just the mark of a concept, and as long as the shape "... is as if it had been constructed like this once and for all (by whoever put the essential properties into things)",[40] it can only be used as a simile.[41]

Furthermore, in face of the suggestion that this essentiality of the figure or the form might exclude any other possibility of partition, Wittgenstein points out that when talking about essence, we merely denote a convention, and that "... to the *depth* that we see in the essence there corresponds the *deep* need for the convention".[42] So, when we say that a proposition expresses the essence of a form, we mean to say that the proposition expresses a property of that form. This is to say: that which is sustained by the proposition and is called 'form' "... is the picture (*Bild*) which I cannot help having when I hear the word 'form'".[43]

The scope of the ideas just discussed, as well as the sense of the word 'picture' that has been used throughout this text, will be made clear if we remember that for Wittgenstein, "A picture (*Bild*) is a fact. What is the case – a fact – is the existence of states of affairs. A state of affairs is a combination of objects (*Gegenständen, Sachen, Dingen*)".[44] And that "We picture facts to ourselves".[45]

Nonetheless, one must not give in to the temptation of showing a determinate object in response to a given picture. Wittgenstein insist: we can imagine a story made from a series of schematic pictures, contrary to one made up of realistic ones, and quite near to our everyday language: it is evident that in order to 'understand' the schematic pictures, there is no need for us to translate them into realistic ones. Just as the coloring of old white and black movies is not needed in order to understand them, even if there are no plants, animals, men, and women in shades of gray.[46] Carrying this argument into mathematics, Wittgenstein will not leave any room for doubts:

> Nothing easier than to imagine a four dimensional cube! It looks like Figure 4. But I don't mean that, I mean something like Figure 5, but with four dimensions. – 'But *isn't* what I showed you in Figure 6, only with four dimensions?' – No; I don't *mean* that. – But what do I mean? What is my picture? Well, it is not the four dimensional cube as you drew it. I have

now for a picture only the *words* and my rejection of anything you can show me.[47]

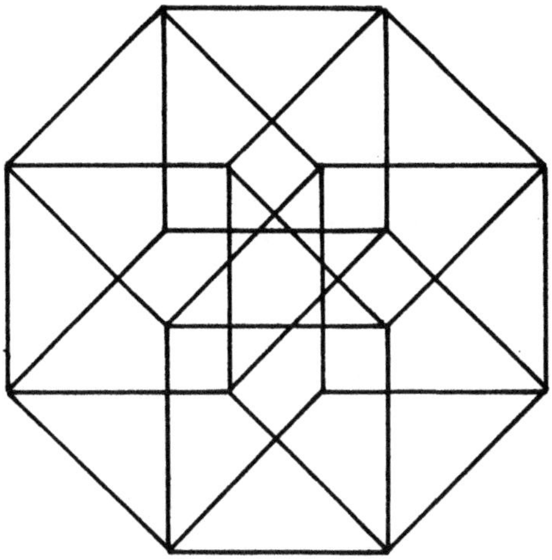

Fig. 4.

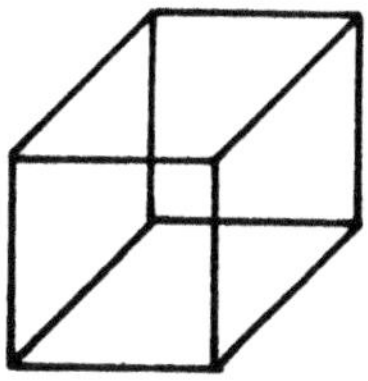

Fig. 5.

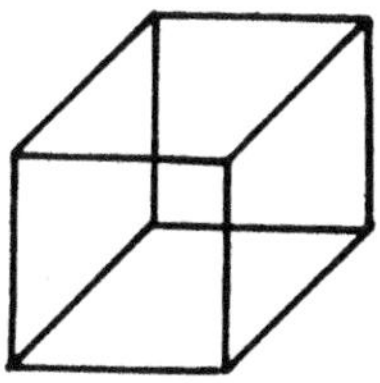

Fig. 6.

Let us examine, in closing, the following remarks.

Once the rectangular figure made up of parallelograms and triangles we have been discussing is accepted as the proof of a proposition, we have (together with the several characteristics of the proof already given), the impossibility of not accepting that, in fact, such a perimeter is obtained with those figures.

If this is the case, Wittgenstein argues that one might think that the matter does not finish with the acceptance of the simpler geometrical proposition, that is, the arrangement given for the rectangle; but it would seem as if we are accepting something beyond a strictly geometrical proposition.[48]

Indeed, even if it is hard to be surprised by the figure, insofar as nothing happens in it, it is yet surprising the way in which straight and skewed go together; it astonishes us that this form has been produced with such figures; this result is amazing after we expected something different,[49] even if we are convinced of the ease with which we can construct the figure from the given parts as soon as we know how to do it.[50]

But, if the picture we have been analyzing (or anything as a matter of fact), is recognized or accepted by someone as a proof, he must accept that in it "*This* follows inexorably from *that*".[51]

We have come round a full circle, we are back to inexorability referred now to proof. But, the figure as proof

> . . . can be 'inexorable' only in its consequence! For otherwise it is nothing but a picture. What does the action at a distance – as it might be called – of this pattern consist in?[52]

Wittgenstein answers himself: "In the fact that I apply it".[53]

Thus, inexorability in a proof comes out through application (or the possibility of doing it, a use) and not through what goes on in the drawing or the figure in themselves.

It is this which allows us to say that in a proof we are in agreement (a notable one) with someone; we go through the proof and accept its result, after all, ". . . this is simply what we do. *This is use and custom among us, or a fact of our natural history*".[54]

NOTES

[1] Wittgenstein, L. *Wittgenstein's Lectures, Cambridge 1930–1932*. Edited by Desmond Lees. Basic Blackwell, Oxford, 1980, p. 24.

[2] Wittgenstein, L. *Vid. Remarks on the Foundations of Mathematics*. G. H. von Wright, R. Rees, and G. E. M. Anscombe (Editors). Translated by G. E. M. Anscombe. The MIT Press, Cambridge, Mass. & London, England, 1978, I.3, p. 36.
[3] *Ibid.*, I.4, p. 37.
[4] *Ibid.* [5] *Remarks*, IV.52, p. 253.
[6] *Remarks*, I.5, p.'38.
[7] *Ibid.*
[8] *Remarks*, I.6, p. 39. Emphasis is mine.
[9] *Remarks*, I.9, p. 40.
[10] *Vid., Remarks*, I.7, p. 40.
[11] *Vid., Remarks*, I.8, p. 40.
[12] *Vid., Remarks*, I.11, p. 41.
[13] *Remarks*, I.10, p. 41.
[14] *Vid., Remarks*, I.15, p. 43.
[15] *Vid., Remarks*, I.16, p. 43.
[16] *Vid., Remarks*, I.17, p. 43.
[17] *Remarks*, I.17, p. 43.
[18] *Vid., Remarks*, I.21, 22, p. 46. We now remark the fact that the concept of Picture (*Bild*) in Wittgenstein's thought is an important one, we shall look into it more closely lately.
[19] *Remarks*, I.33, p. 50.
[20] *Remarks*, I.25, p. 46.
[21] *Remarks*, I.26, p. 47.
[22] *Vid., Remarks*, I.27, p. 48.
[23] It is quite instructive to compare, here, what S. Kleene has to say on proof: "The proof of theorems, or the deduction of consequences of assumptions, in mathematics typically proceeds à la Euclid, by putting sentences in a list called 'proof' or 'deduction'. We use the word 'proof' when the assumptions have a permanent status for a theory under consideration . . . " Kleene, S. C., *Mathematical Logic*. Wiley & Sons, Inc., New York, 1967, p. 33.
[24] *Vid., Remarks*, I.28–I.30, pp. 48–49.
[25] *Vid., Remarks*, I.31, p. 49.
[26] *Vid., Remarks*, I.34, p. 50.
[27] *Remarks*, I.32, p. 50. Emphasis is mine.
[28] *Remarks*, I.73, p. 64.
[29] *Remarks*, I.36, p. 51.
[30] *Vid.*, Cantor, G.: *Contributions to the Founding of the Theory of Transfinite Numbers*. Dover Publications, Inc., New York, 1955.
[31] *Vid., Remarks*, I.42, p. 55.
[32] *Vid., Remarks*, I.43, p. 55.
[33] *Remarks*, I.44, p. 56.
[34] *Remarks*, I.50, p. 57.
[35] *Ibid.*
[36] *Remarks*, I.69, p. 63.
[37] *Remarks*, I.68, p. 63.
[38] *Remarks*, I.71, p. 63.

[39] *Remarks*, I.72, p. 64.
[40] *Ibid.*
[41] *Ibid.*
[42] *Remarks*, I.74, p. 65.
[43] *Ibid.*
[44] *Vid.*, Wittgenstein, L., *Tractatus Logico-Philosophicus*, translated by D.F. Pears & B.F. McGuiness. Routledge & Kegan Paul, London and Henley, 1981, pp. 5–9.
[45] *Ibid.*
[46] *Vid.*, Wittgenstein, L., *Zettel*. G.E.M. Anscombe & G. H. von Wright (eds.). Translated by G.E.M. Anscombe. University of California Press, Berkeley, 1970. Paragraphs 241, 242, p. 44e.
[47] *Zettel*, paragraph 249, pp. 46e, 47e.
[48] *Vid.*, *Remarks*, I.57, p. 59.
[49] *Vid.*, *Remarks*, I.58, I.60, p. 60.
[50] *Vid.*, *Remarks*, I.59, p. 60.
[51] *Remarks*, I.61, p. 60.
[52] *Remarks*, I.62, p. 61.
[53] *Remarks*, I.65, p. 62.
[54] *Remarks*, I.63, p. 61.

CARLOS TORRES AND JAIME OSCAR FALCON VEGA

TO SHOW AND TO PROVE*

It is often said that mathematics are a deductive science. Inasmuch as this claim puts forward its demonstrative character, it has two failures: first, it does not account for the creative or heuristic side of mathematics (for example, how are theorems discovered?); second, it omits the fact that a great part of mathematical knowledge is based upon sense evidence, not on proof. With this we mean that there are – in the last instance, for in practice the process is quite mixed – two kinds of 'truth' in mathematics: those based on sense-perception and those inferred from other propositions by logical deduction. Those, we say, are *shown*, the latter are *proven*.

The purpose of this paper is to explore the notions of showing and proving, and to try to establish the need for proof in mathematics. With such an aim, we look at Pythagorean mathematics since it was in its midst that both tendencies flourished (the visual and the deductive) and the first mathematical proof was produced in the modern sense of the term.

1. SEEING IS BELIEVING

Observation and evidence are the primary sources of mathematical knowledge. At the beginning a truth was discovered through direct inspection of figures. *Let us see* Pythagoras theorem in Figures 1, 2, 3 and 4.

Everyone knows the Pythagorean theorem. It states that the square constructed on the hypotenuse of a right triangle is equivalent to the sum of the squares built upon the two other sides. *As we have seen*, the proposition is true. The same is shown in the following figures (Figure 5) that, according to tradition, were known to Pythagoras.

In showing this, it is assumed that rearranging the triangles in the big square does not alter them. It presupposes the notion that while moving a body from one place to another, it remains the same geometrical body, that the properties of geometrical figures are the same regardless of

S. Ramirez and R. S. Cohen (eds.), Mexican Studies in the History and Philosophy of Science, 249–264.

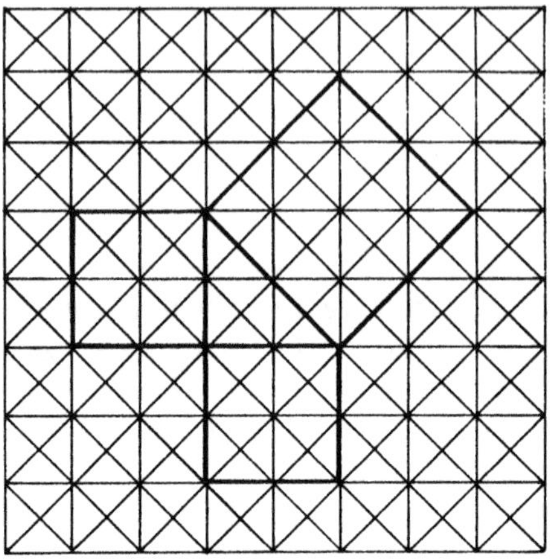

Fig. 1. Arabian tile.

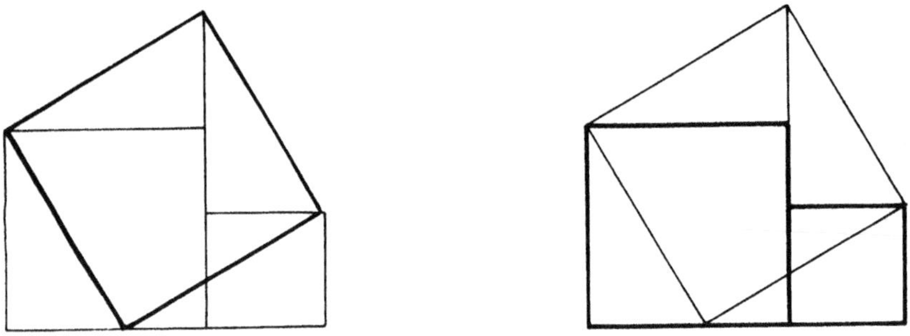

Fig. 2. H. Perigal 1873.

position. But that is the way the figures are meant to be. Something analogous can be said about the Bhaskara figure.

2. NUMBERS AND FIGURES

Everything seems to indicate that Pythagoras was led to the discovery of his theorem by verifying that the triangle with sides 3, 4, 5 has a right angle. One way of doing it is by using a square of side 7 (= 3 + 4) and dividing it into 49 square units. We show this figure (Figure 6) divided in two different ways.

As one can *see*, the square of side 7 can be divided in:

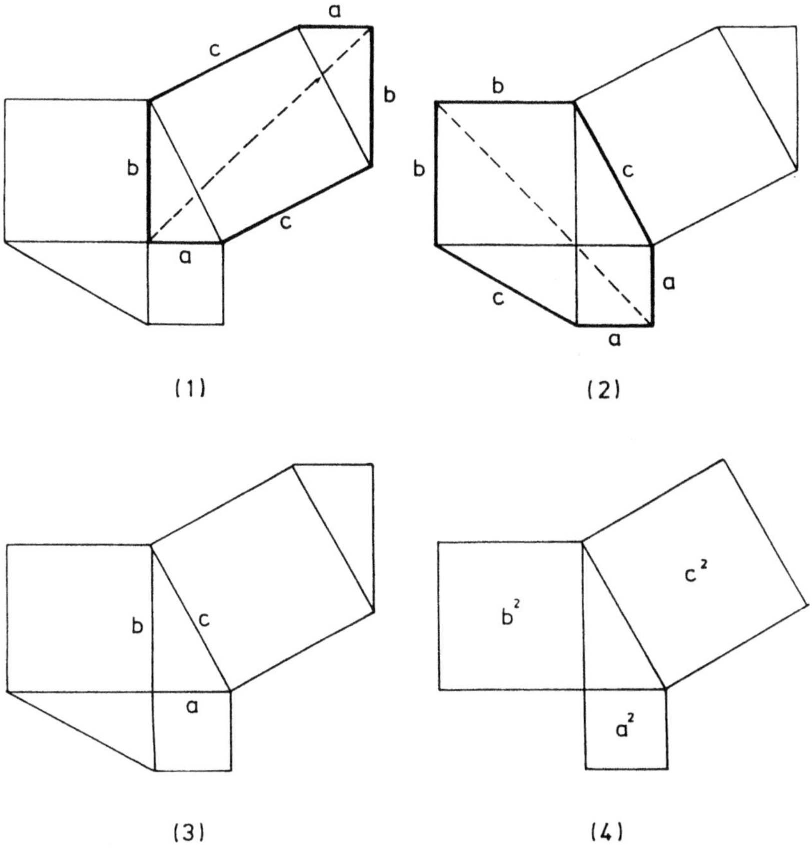

Fig. 3. Leonardo da Vinci (1452–1519).

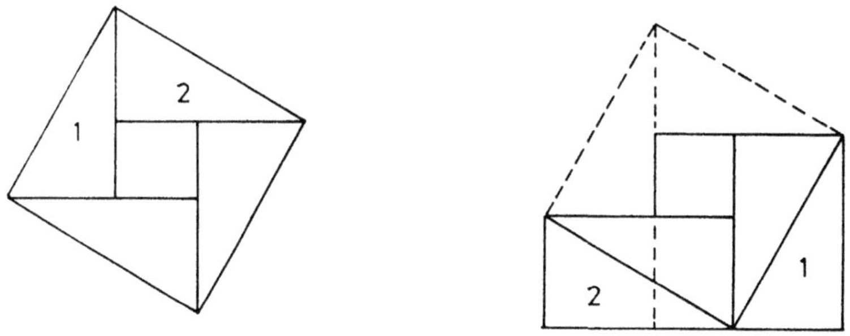

Fig. 4. Bhaskara 1114 a.c.

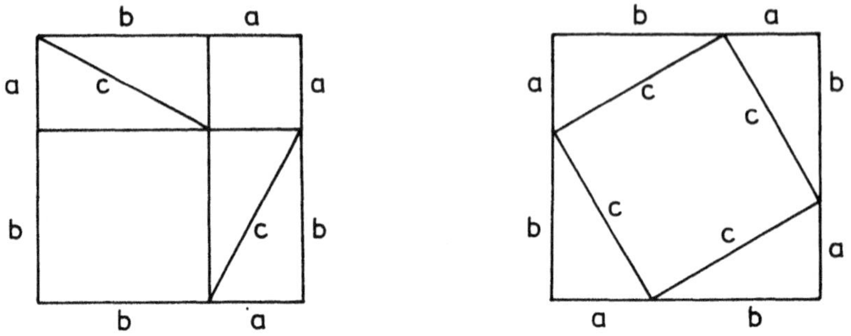

$a^2 + b^2 = (a + b)^2 - 4\frac{a \cdot b}{2}$ $\qquad$ $c^2 = (a + b)^2 - f\frac{a \cdot b}{2}$

$$a^2 + b^2 = c^2$$

Fig. 5.

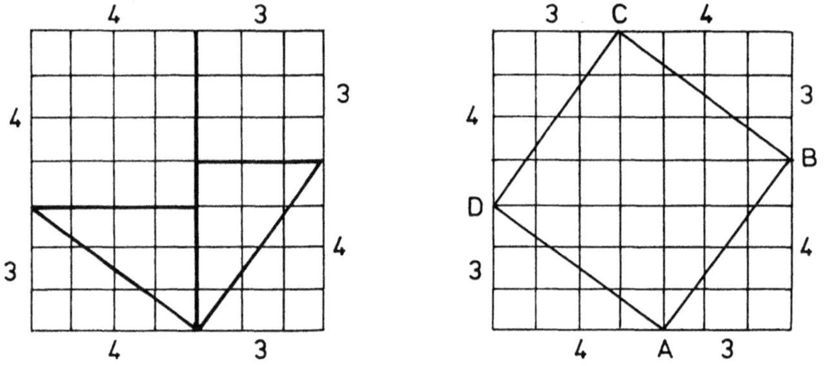

Fig. 6.

(a) four halves of the rectangle of sides 3 and 4 and the squares 3^2 and 4^2;

(b) four halves of the rectangle 3, 4 and the square ABDC;

whence the square ABCD is equal to the square 5^2 $(= 3^2 + 4^2)$ and its sides are of length 5. This shows that the triangle of sides 3, 4, 5 is a right triangle.

This discovery would lead him to search in other similar cases. The method used can easily be extended when the sides of the triangle are

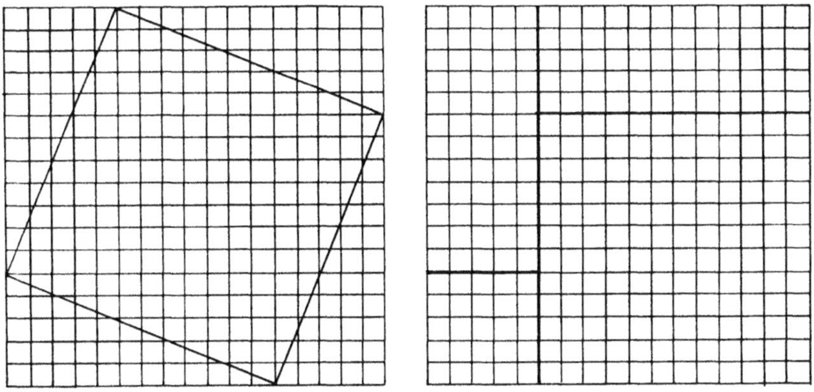

Fig. 7.

natural numbers a, b, c such that $a^2 + b^2 = c^2$. The general idea is the following: a square of side $a + b$ can be divided into $(a + b)^2$ smaller squares. When this is done, it can be subdivided in two different ways:

(a) as two rectangles with sides a, b and squares a^2 and b^2,
(b) as four halves of the rectangle of sides a, b and the square on its diagonal.

When these two partitions are compared, the square on the diagonal of the rectangle a, b will be equivalent to the sum of the squares a^2 and b^2. As this sum is c^2, the diagonal of the rectangle of sides a, b has length c and the triangle of sides a, b, c is a right triangle.

For example, $5^2 + 12^2 = 13^2$. The triangle of sides 5, 12, 13 is a right triangle as the figures in Figure 7 show.

As it can be seen, this procedure is essentially the same as the one attributed to Pythagoras, of which it is but a special case: that in which the sides a, b, c are integers and the sum of its squares another square.

Pythagoras theorem has as a consequence that all numbers satisfying the relation $a^2 + b^2 = c^2$ produce a right triangle. Thus the notion of perpendicularity is transformed to numerical terms; every time the numerical relation takes place, the geometrical one also takes place; as if numbers were an expression of the harmony between things and things obeyed the numerical laws.[1]

This relation leads us to investigate the *arithmetical* problem of finding integer numbers that would be the sides of a right triangle, or equiv-

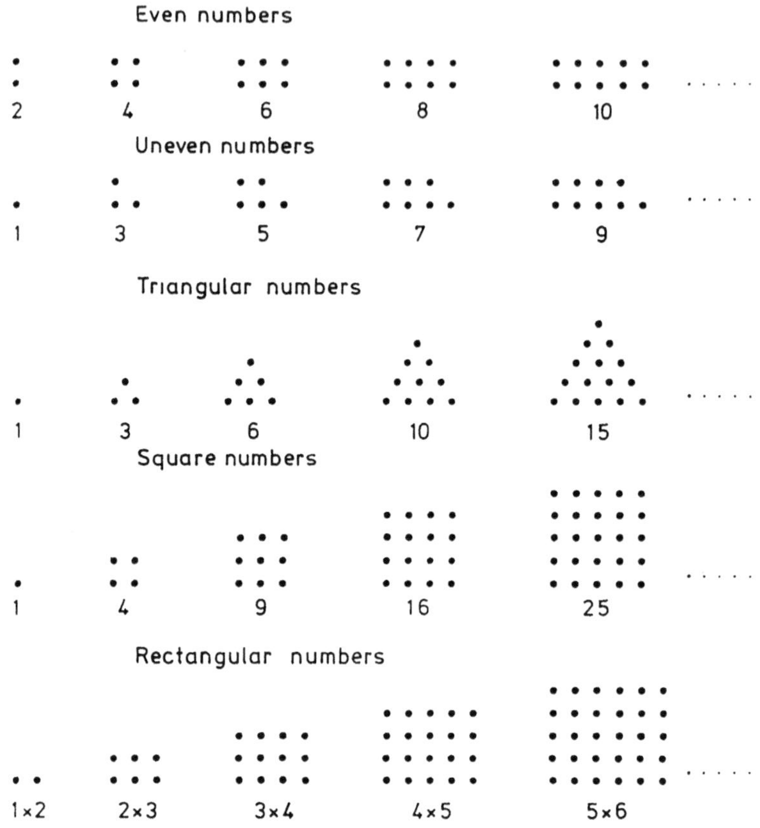

Fig. 8.

alently, to find square integers that are the sum of two squares. While doing so, we shall not argue from general principles, but from direct observation of the figures representing numbers (*figured numbers*).

A *figured number* is a number that can be represented in geometrical form by using points. The notion is shown in Figure 8.

Many properties are readily seen in these configurations. Behold (Figure 9a):

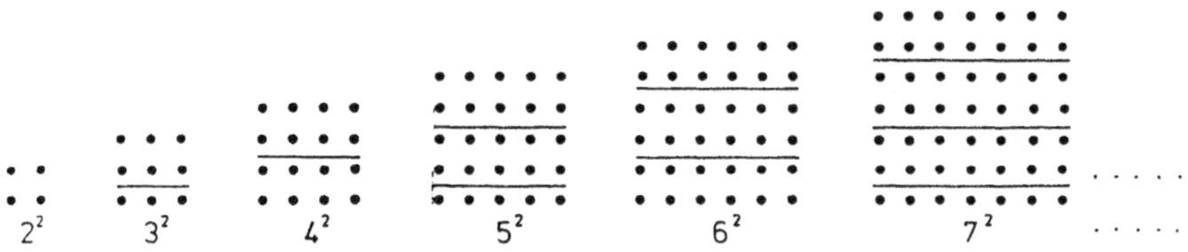

Fig. 9a.

The square of an even number is even; the square of an uneven number is uneven. Squares are as their roots, even or uneven (Figure 9b).

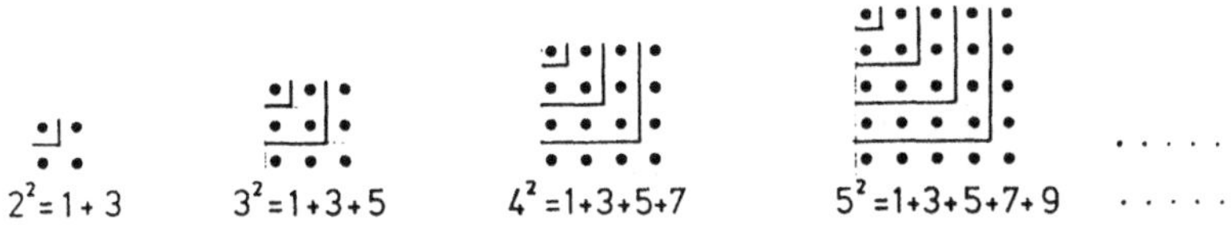

Fig. 9b.

Every square number is the sum of all the odd numbers that are less than the double of their side (the root). Square numbers grow as the uneven numbers grow (see Figure 9c):

$$
\begin{aligned}
n^2 &= 1 + 3 + 5 + \cdots + (2n - 1); \\
(n + 1)^2 &= 1 + 3 + 5 + \cdots + (2n + 1).
\end{aligned}
$$

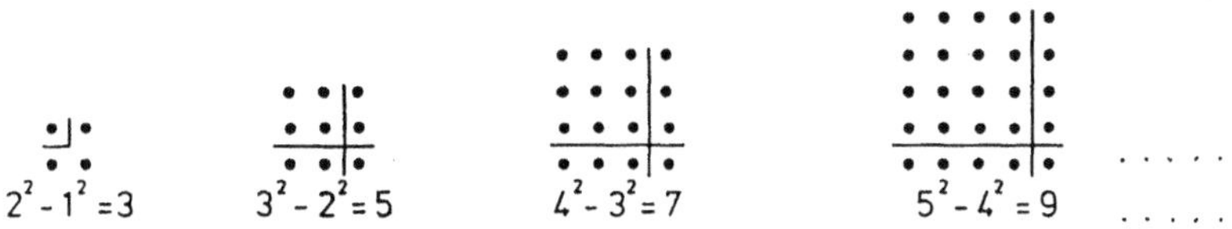

Fig. 9c.

The difference between two successive squares is uneven: $(n+1)^2 - n^2 = 2n + 1$ or, $(n+1)^2 = n^2 + 2n + 1$.

The formula $(n+1)^2 - n^2 = 2n + 1$ states a relation between square and uneven numbers. It states that the sequence of uneven numbers is the sequence of differences between consecutive squares. Let us write such a relation in a table:

natural:	1	2	3	4	5	6	7	8	9	10	11
squares:	1	4	9	16	25	36	49	64	91	100	121
uneven:	1	3	5	7	9	11	13	15	17	19	21

12	13	14	15	16	17	18	19	20	21	22	
144	169	196	225	256	289	324	361	400	441	484	
23	25	27	29	31	33	35	37	39	41	43	45

The uneven numbers are the difference of the squares immediately above them. In order to find square numbers whose sum is also a square, it will be enough to pick, from last line in the table above, those numbers that are squares. Thus, by simple inspection, one can find integer numbers that are the sides of a right triangle. For example $9 = 3^2$ and the triangle with sides 3, 4, 5 is right; just as $25 = 5^2$ generates the triangle 5, 12, 13, etc.

Knowing the rule, it is natural to look for a formula that allows us to find the triads of numbers a, b, c that belong to the sides of right triangles. Uneven numbers are those of the form $2n + 1$ and uneven squares satisfy $2n + 1 = m^2$. If we assume that the latter is satisfied, then

$$n = \frac{1}{2}(m^2 - 1) \quad \text{and} \quad n + 1 = \frac{1}{2}(m^2 + 1).$$

In order for n and $n+1$ be integers, m must be uneven. If we consider that $n^2 + (2n+1) = (n+1)^2$ we get the formula:

$$\left|\frac{m^2 - 1}{2}\right|^2 + m^2 = \left|\frac{m^2 + 1}{2}\right|^2 \qquad (m \text{ uneven}).$$

This formula corresponds to what is now known as the Pythagorean rule. All the numbers given by this formula give rise to the geometric relation of perpendicularity (see Figure 10).

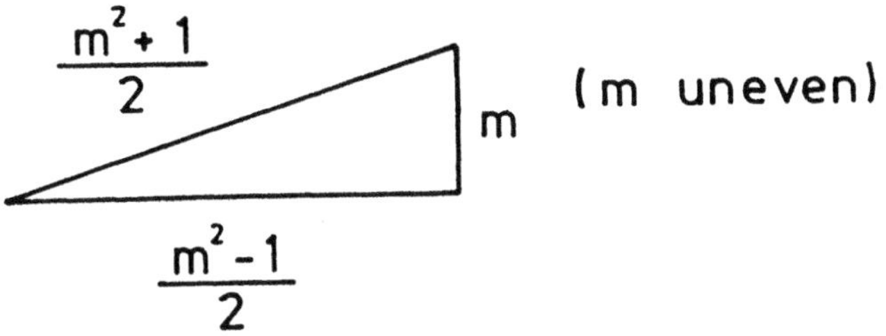

Fig. 10.

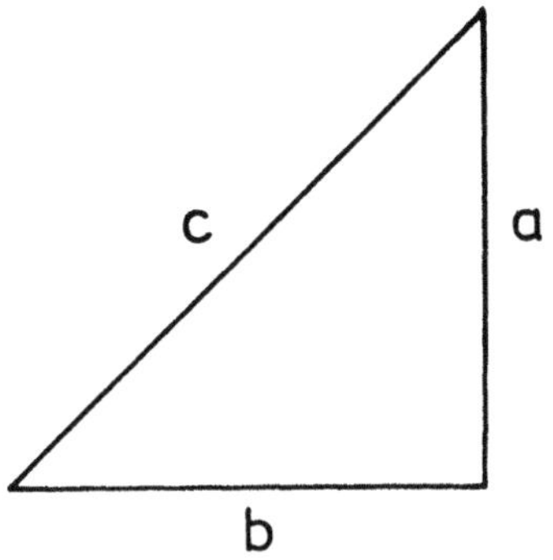

Fig. 11.

For example, when the values 3, 7, 11 and 15 are given to m, the Pythagorean triads 4, 3, 5; 24, 7, 25; 60, 11, 61;, 112, 15, 113 are determined and are called *Pythagorean numbers*.[4]

This formula does not give all the Pythagorean triads (it only yields those for which $b < a = c + 1$). This limitation to its scope shows the need to explore the cases it does not contemplate (e.g., 8, 15, 17; 29, 33, 45; 48, 55, 73; etc.).

3. WHAT DOES NOT EXIST CANNOT BE SHOWN

A particular case of the problem is the following: *to find natural numbers that are the sides of a right isosceles triangle*. Once we have Pythagoras theorem it is natural to consider this problem under the light of the numerical relations it involves (see Figure 11) and we have the relation

$$a^2 + a^2 = c^2$$

that is

$$2a^2 = c^2.$$

It is a very simple relation: the square of c is two times the square of a; a^2 and c^2 are two squares such that one is twice the other. The arithmetical problem is equally simple: to find a square number that is twice another square.

To cope with the difficulty we have a method that has been proven efficient in the general case. Let us write three columns (a) of natural numbers, (b) their squares and (c) the double of these. To solve the problem we have to choose, from the third column, those numbers that are squares (see Table 1).[5]

In what we have written there are no squares in column (c). When looking for it to appear, it might never appear. If this were the case, our table will never show it: our search if finite and the table is infinite. Let us consider the situation. While inspecting the table without finding a square in column (c), is it because we have not gone far enough? How can we know if there are no such numbers? Our method can only show a number when such a number exists, but it is incapable of providing answers when the number does not exist. How can this problem be solved? *Let us think about it.*

Let us *imagine* that we have found the first square on column (c), say c^2. As it is the double of another square, we have

$$c^2 = 2a^2$$

as it is an even number,

$$c = 2b$$

then

$$2a^2 = 4b^2$$

that is

$$a^2 = 2b^2.$$

But as $2b^2 < 4b^2$, $a^2 < c^2$. This states that a^2 comes before c^2 in the list (c). Hence, before getting a first square being the double of another one, we must have previously encountered another! This shows that the existence of such a number is impossible: while looking for a square that is the double of a square it will never appear.

We have *proved* that:

TABLE 1.

(a)	(b)	(c)
1	1	2
2	4	8
3	9	18
4	16	32
5	25	50
6	36	72
7	49	98
8	64	128
9	81	162
10	100	200
11	121	242
12	144	288
13	169	338
14	196	392
15	225	450
16	256	512
17	289	578
18	324	648
19	361	722
20	400	800
21	441	882
22	484	968
23	529	1058
24	576	1152
25	625	1250
26	679	1352
27	729	1458
28	784	1568
29	841	1682
30	900	1800
31	961	1922
32	1024	2048
33	1089	2178
34	1156	2312
35	1225	2450
36	1296	2592
37	1369	2738
38	1444	2888
39	1521	3042
40	1600	3200
41	1681	3362
...	...	...

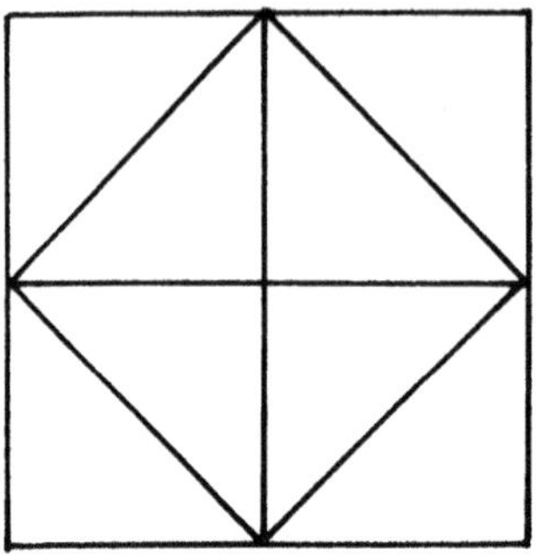

Fig. 12. Pythagoras theorem in its simpler case: the right isosceles triangle. The truth of the theorem is manifested with the figure. It is shown, it is evident.

There is no square number that is the double of a square,

or

There are no right isosceles triangles with integer sides.

This is a kind of knowledge to which we have arrived not by inspection of figures or through sense-evidence: what does not exist cannot be shown. It is only through an argument that we can know about the non-existence of a square number. This justifies the need for proof in mathematics. Without it, we could never know these facts. It is only through proof that we know that which cannot be shown, that which transcends what is shown.

4. TO SHOW AND TO PROVE

To show and to prove are not different in their principle: what is proven is shown and we prove that what we show. At this level, we take as true everything our senses say is so. The foundation of truth, in what we have done, is: we say what we see is in the figure. The figure occupies, here, a dominant place and becomes a proposition. As a pictogram, it points beyond itself as if it were saying: "here is the truth, you have reached if" (Figure 12).

Through sense-evidence we have established the Pythagorean theorem or the truth of the proposition: "every uneven number is the difference of two consecutive squares". It is enough to look closely at the figure to state that what has been said is true. This side of mathematics has been sufficiently emphasized.

At a more sophisticated stage, proof is clearly differentiated from perception and sensibility as a way to truth. When confronted with the impossibility of showing the non-existence, the need for proof arises.

We have made the need for proof in mathematics conspicuous. Now the question: what are the grounds for proof? Why do we take as true that which has been proved? Let us look at the only proof we have considered, the proof of the proposition:

"There is no square number that doubles a square"

We assumed the opposite, namely,

"There is a square number doubling a square"

and we considered the smallest number with this property by giving it a name, c^2. According to our assumptions, the following were true:

(1) There is a number a such that $c^2 = 2a^2$.
(2) There are no numbers x, z such that $x^2 < c^2$ and $c^2 = 2z^2$.

After (1) we inferred that there was a number b such that $a^2 = 2b^2$ which led us to:

(1′) $a^2 < c^2$ and $a^2 = 2b^2$.
(2′) There are two numbers x, z such that $x^2 < c^2$ and $x^2 = 2z^2$ (e.g. $x = a$ and $z = b$).

As the last statement is the opposite of (2) it contradicts it.

We conclude that the statement "there is a square number doubling a square" is not true since contradictory propositions follow. Formally, the rejection is based upon the logical principle of non-contradiction whose validity was implicit in the argument. Accordingly, the two propositions, being contradictory, cannot be simultaneously true. The principle can be stated as follows:

"not (E and not E)".

In the case we are dealing with, from the assumption "there is a square number doubling a square" we inferred that the statements

"there are no numbers x, z such that $x^2 < c^2$ and $x^2 = 2z^2$"

and

"there are numbers x, z such that $x^2 < c^2$ and $x^2 = 2z^2$"

are true, and this dissents from our principle of non-contradiction. As the assumption is not true, its negation, namely, "there is no square doubling a square" is true. Formally, the admission of this fact relies on the logical principle of the excluded middle. According to it, when two statements contradict each other they cannot be simultaneously false. It is enough, then, that the falsity of one is recognized to assert the truth of the other. This principle can be stated as follows:

"E or not E".

In our case, "there is a square number that doubles a square or there is no square number doubling a square". As the first one is not true, the second one is.

This is how the proof of the statement "there is no square number doubling a square" was carried out.

In its most common form of acceptance, *to prove* is to deduce from principles or premises taken as truths.

What is thus the ground for proof? Certain logical principles as non-contradiction or excluded middle.

Why do we accept as true what has been proved? Because from these principles we have the certitude that it cannot be otherwise. In this sense, rational necessity is the certitude of deduction and it must be argued as such. On the contrary, a truth that is grounded and sense-evidence requires no argument since the presence of the fact or of the object makes it clear.

Certitude and evidence; reason and senses: such are the modes of acquiring truth in mathematics. Certitude is subjective security about the truth of knowledge; evidence is the way in which things offer themselves to the senses in order to be 'understood'. From this point of view, evidence is not a subjective matter but rather an objective one, it is not linked to clarity of ideas but to the presentation and manifestation of objects. Between these two extremes, knowledge in mathematics arises.

NOTES

* [This text was originated at a conference for high-school teachers of mathematics delivered by Jaime Oscar Falcón Vega. After lengthy conversations between the authors,

Carlos Torres extended the analysis of the role of proof in mathematics and wrote the present text. When it was finished and submitted to the editor, Jaime Oscar Falcón Vega had the unexpected opportunity of reading it and discussing anew with Carlos Torres. This produced the post-scriptum – to which both authors adhere (for the time being) – S.R.]

[1] This must have been immensely satisfactory for Pythagoreans who, or so it seems, came to believe that the principles of mathematics were the principles of all existing things. The discovery that musical harmony depends on numerical proportions (the octave represents the proportion 2:1 of the length of the string, the fifth the proportion 3:2, etc.) must have produced a very profound impression reinforcing the idea that numbers allows things to be intelligible. This theory appears in many ways. Philolaus says that *all things have numbers* and that this fact makes it possible for them to be known. A more radical form of this theory is that claiming that *all things are numbers*.

[2] The word in Latin is *par, paris*, 'equal'. Even numbers are those whose points can be paired.

[3] Everything indicates that at the dawn of Greek culture, numbers were represented with points. Later, when different configurations were formed, figured numbers must have appeared. This would be the origin of number theory, where geometric intuition plays a remarkable role. Today the geometric origin can be found in such words as *cube* and *square*.

[4] *Pythagorean triad*, a, b, c such that $a^2 + b^2 = c^2$.

[5] For typographical reasons we change rows for columns.

POST-SCRIPTUM

How is sense produced? We do not really see Pythagoras's theorem through senses. What we see in the Arabian tile is a tile. To show is not only related to the sensory way. It is rather a way of incorporating the figure into language. What we must solve is the manner in which the figure works as a proposition, how it operates as a part of language.

What does Pythagoras's theorem state? "The squares on the sides of the right triangle build conjointly the square on the hypotenuse". The content of this statement is precisely the figures showing the statement.

Every even square consists of four squares. See:

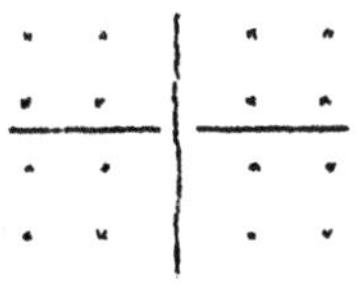

If a square doubles a square, the latter will double the four squares of the former square. So that your search in the list is bound to fail. The first square that doubles a square cannot be the first one, the doubled square would be the first one, because it doubles the four squares in the former. The figure is working for logic and is not logic.

Logic is already prejudiced in the geometrical objects. Just as order is prejudiced in the list (c) (and just as the Euclidean geometry of the plane is prejudiced in the figures).

The principles of logic do not have the character of universal laws. When I argue, and this is clearly against Russell's position which reconstructs logic before reconstructing mathematics, logical principles are not empty skeletons that, when filled produce a truth. That is why Hilbert wanted a simultaneous reconstruction of logic and mathematics within a formal system.

It is remarkable that the endless search in the list (c) led to a proposition about the infinite list, namely "there are no squares in (c)". The mathematical interest is shifted to the grammar of infinity, to the question "how is it possible to construct a discourse about infinity?"

SANTIAGO RAMIREZ CASTAÑEDA*

A RETURN TO VIENNA**

To the memory of Guillermo Torres who bestowed upon me a passion and its risks

"Behold, this dreamer cometh"
(Genesis 37:19)

All atriums hold the scent of the alluring and the seductive texture of the unknown. Trying to pierce into the future is a compelling temptation; to probe into our hopes is what makes us part of the human species. But doorways also harbor our memories and impel us to think again about what we leave behind.

Thus, as we enter the third millennium, we behold a multiplication of calls to bestow our expectations. Yet, it is equally important to try to answer the questions about what we have lived, to answer the mysteries about our own past.

In an effort to provide a reply we can state that, undoubtedly, the most important single fact or at least the most extraordinary event in the life of the ordinary people of the 20th century has been the October Revolution. At the same time, it is obvious that the milestone of our generation, the event that has sealed the fate of our generation most profoundly, is the so-called 'crisis of Marxism' and the ensuing dismantling of the socialist block.

It is also highly likely that this phenomenon has peculiar features that leave it thoroughly unique: the universality of the hopes it generated and the alacrity of their waning. In the lapse of a single lifetime we have experienced what is possibly the most ambitious illusion of the human being and we have also seen its decline with as much awe as the trust we had laid upon it. It might also be possible that our generation is one of those chosen generations of men that has experienced, in such a global manner and in such a brief time, the rise and fall of so many and such unlimited illusions. In one life time we have concentrated what history has wisely and patiently distributed along the ages. It was an illusion as meteoric as its fading. In one, and a fleeting, instant in the infinite

S. Ramirez and R. S. Cohen (eds.), Mexican Studies in the History and Philosophy of Science, 265–273.

history of eternal return, we were allowed to trek along the unlimited circle of hope and despair. It does not lack a dramatic charm: we have not only lived hastily, we also know that many men and women in many places have died heroically to give birth to those specters; they died as Althusser recalled, for a cause that we know now was nothing more than an enormous deception, but men confined in the Gulag wept when Stalin died.

The hallucination that so enchanted us, that made us believe that the true history of humanity was about to begin and that this vast humanity had *said* enough and had started its heroic journey; and sudden vanishing of that hallucination is not unique in our century; there were others, perhaps more mercurial, more violent, less universal; there were others that far from generalizing a shining future for men, showed us the horrible face of the most unblemished evil. Pico della Mirandola had written that men were as close to God as they were close to the most brutal bestialitity. We have lived through all of it as the subjects of unequaled desires and we have attested to the cruelest rage and an unparalleled frustration. We watched the madness of the Third Reich and the terrifying end of World War II which has allowed us a long period of fragile and relative peace whose sole foundation is fear. Man in the 20th century can be thus described, as the being who at the same time harbored an unshakeable belief in the most magnificent dreams and who lived through the ghastliest fear.

All these events, markings of our immediate history, seem to flow through Vienna: through a Vienna that endured, in less than a century, the complete spectrum of human passion: from the melancholy of the Blue Danube to the turbulence of the Red Revolution.

It was precisely at this vortex that the most relevant currents of politics and thought entangled, that the soundest and most controversial issues of thought and culture of our times arose. It was conceivably at Vienna that the ideas that allowed us to bear the end of innocence, fear, and the Holocaust, were generated; but it is also possible that the ideas that made that terror possible were also born at Vienna. It was, thus, at Vienna that the most enduring visions about the most divine features of mankind were built and it was also at Vienna that the most dreadful nightmares about human monstrosity were hatched.

The downfall of the Empire – a shred of the Carolingian dream – was not the effect of isolated events: Sarajevo and Mayerling were but anecdotes in a much more dire history. The decadence was the harvest of the simultaneous action of a throng of conflicting forces that had weakened the structure of what had been the post-Napoleonic peace upon which we would have wished to perpetuate modern thought but upon which a new rationality was dawning.

In this atmosphere of imminent disintegration and unheard-of creativity – brightly described by Karl Kraus or by student Torless – flourished the most rejuvenating and vital forces of thought and imagination of our times. Forces which we have attempted to ignore, or that we have hastily disqualified in the name of the proletariat, or of a romanticism that had already lost its gigantically tragic and powerful dimensions.

It was at Vienna where the ideas that shape our outlook on the cosmos arose; it was at Vienna that the aesthetic currents that have modified our sensibility were cradled; it was at Vienna that the theories that have reshaped the structure of man were outlined: science, art and psychoanalysis; Mach, Mahler and Freud. Knowledge, imagination and passion found new reasons for being, reasons that we, now, can think anew to find ourselves again.

This could be an opportunity to speak about science, art or psychoanalysis. Rather it is more fitted to vindicate the thought of those who were the source of our passion. In any case, this is the time for vindicating the unharnessed and amazing imagination of the men at Vienna.

It is an opportunity to reconsider two apparently secondary questions whose repercussions are much more meaningful than we have heretofore recognized: the issues of mathematics and of philosophy of science; of Gödel and Carnap. Both protagonists of an impossible enterprise: the Vienna Circle; both misunderstood or condemned for the sake of fleeting fantasies. Yet, to vindicate the role of the Vienna Circle does not imply the dismissal of other contributions; it purports rather, reconsideration of a tradition of which we have been deprived.

HILBERT'S PROPHECY

From the point of view of mathematics we can pinpoint a radiant moment: Hilbert takes the floor in the 1900 Congress of Mathematicians and proposes the now famous list of the problems that, still today, are the axis of modern mathematics. Not only a problematic axis but also the focal point around which mathematical knowledge and mathematical practice have been reconstructed.

If we accept that a scientific theory is the result of a peculiar interlocking of concepts, we must also admit that the central concepts that the Hilbertian program weaves in an extraordinarily original way are those of invariant, transformation, the infinite, and continuity. Hilbert, of course, did not invent them, he simply reorganized them. The relation had already been suggested by equally articulate mathematicians: Dedekind and Klein, Cantor and Bolzano. But, as Marx said about Hegel, Hilbert knows what the others simply do.

Moreover, if we accept that concepts are answers to problems, and that the entanglement of concepts becomes a theory, we must admit that concepts lay down the architectural blueprint that organizes theory and they are the consequence of outlining a problematic perspective. This horizon is what Hilbert sketched and, while doing it, he placed mathematics in a different and utterly original situation. It is in this quarry unearthed by Hilbert, that Gödel's true task takes place. Furthermore, it could not have any sense elsewhere.

Hilbert again astounded the mathematical world with his work on the infinite: it is the declaration of mathematical independence; if Cantor had created for us – mathematicians – a paradise, Hilbert had the lucidity and eloquence needed to provide us with consciousness of it.

In this paradise, where mathematics grows as an absolute entity, the cravings of philosophy, by which external criteria were to be admitted in mathematics, are repealed. Mathematics becomes the supreme court of mathematical thought; mathematics can thus ignore a previous logic and a prior metaphysics and it can exclude any ensuing criteria to construct, reconstruct or propose its truth. Moreover, this idea of truth itself becomes a matter of mathematical thought alone.

As in every paradise, conflict and rebellion appear, false prophets and false creators, golden calves and towers of Babel. I will not describe here and now the progression of these celestial uprisings and the intestinal wars that took place among the manifold attempted explanations. It is enough to mention that the battlefields and that the rules of the game were, and possibly still are, those that Hilbert displayed. It seems appropriate, nevertheless, to describe the battles that were spearheaded – with its variations and its hesitations – by the Vienna Circle.

PARADISE LOST

It is a well-known fact that Carnap attributed the intellectual origin of the Vienna Circle to Russell, Einstein, and oddly enough, to Wittgenstein.

But there are other undeclared forefathers: Hilbert, Kraus and Freud. Above all, the Vienna Circle was made possible by a new way of thinking, the ancient problem of language.

For the Vienna Circle is undoubtedly constructed upon a specific development of the notion of language, upon an array of promises that had been formulated in other contexts, upon new ways of placing and attacking those problems that since Cratilus and St. Augustine had not been reconsidered under a new light. Underlying the Circle, a revolutionary view of language and of logic were necessary.

This view was, briefly, that which guaranteed the possibility of reconstructing thought upon a unified base, upon a unique logic that underlaid every possible thought, and upon a language that could be liberated from its ancestral and burdensome ambiguity.

It is a view of the world by which what is said coincides with being, by which being is reflected unequivocally onto language. This is the mortal sin, the greatest illusion, and the impossible task of Carnap.

This heroic task – which must be measured in terms of its goals and not in terms of its achievements – began with the overhauling of language itself and with the development of a new syntax. This is Carnap's road to heaven, and his legacy cannot be judged in terms of where he led us but rather in terms of the force of his conviction and in terms of the faith with which he endowed human endeavor; in his uncompromising search for a truth to be pursued even when we know that it will never be revealed.

We were aroused from the Carnapian dream by Wittgenstein: *malgré tout* there is a philosophy of values contrary to a philosophy of knowledge, those obstinate and persistent forms of metaphysics – as Bachelard calls them – that reappear constantly, always renewed, always new ... Nevertheless Carnap's step was a radical one: to claim that metaphysics must be abolished is a radically new way of thinking metaphysics, of thinking about the same but differently.

The dismissal of Carnap's project may be valid for philosophers. But Carnap's project is perfectly just if we judge it from the point of view of Hilbertian mathematics. Mathematics – and this was Hilbert's requirement – has parted with every prior metaphysics in order to be capable of thinking by itself and it is the only representative of thought that does not need the 'removal of ambiguity' because fuzziness is eliminated by mathematical practice alone. After all, mathematics takes care of itself. Every confusion originates in semantics and Hilbert's mathematics – as well as Carnapian mathematics – is exclusively syntactic. To renounce such a syntactic view is to nullify Cantor's paradise.

In this context formalism acquires a first rank importance and with it, with Hilbert, the problem of infinity can be posed in its unfathomed magnitude.

Success is spectacular; almost all formalized arithmetic is complete. It only remained to prove that the introduction of the product will not alter the consistency of the reconstructed theory. The main actor of this optimism is a minor member of the Circle, Karl Gödel, but Gödel provoked a crisis; the crisis he unleashed can only be compared with the Pythagorean crisis: without a shadow of doubt and avoiding every logical, metaphysical or methodological objection, Gödel's theorems, overwhelmingly admitted by all, become the source of a legion of catastrophic and pessimistic interpretations about human thought and about the being of man in the world; they are taken as a mortal blow to the role of mathematical thought, and against the sought for structure of language; they lead to apocalyptic speculations of almost every lineage.

Nevertheless, the theorems confirm in the most radical mode the impossible position of a form of thought that becomes human because of this impossibility. Fifty years later we can state that Gödel's theorems are the vindication of our condition, of the irreplaceable position of mathematics at the core of that which exists as 'the most radiant beauty' and the uncontested proof of the existence of the Cantorian paradise.

THE DEATH OF DESCARTES

French thought has never recovered from the death of Descartes – almost every French philosopher has tried, in his own peculiar manner to be the overseer of thought that is basically unrepeatable. Year after year, France discovers new reincarnations of Descartes, and every year they disappear with the same swiftness by which they were created. Few survive their own fame: Bachelard, permanently generous, and Cavaillès, constantly surprising.

I think about these two French philosophers because they are our tradition. A tradition we have built boldly. I think about Bachelard and

Cavaillès because it has been in terms of them that some of us – but only some of us – have been able to go beyond the humiliating role of repeating, and have had the insolence of trying to be original. We have often been deprived, quoted, used and corroded, we have been accused of almost everything and we have seldom responded, but our silence – this nonsense by which we do not engage in apparently decisive battles, an indifference in fact of the surrounding mediocrity – is the same with which Cavaillès faced the firing squad. Because we share the same passion: I will not accept to confine my imagination to the obscure dungeons where I might only mourn the loss of a world which I had already forsaken, I will not allow myself to nourish the desire for a paradise that cannot be experienced except as paradise lost: I will not forfeit the world of my impudent imagination, of my disrespectful brashness: my hyperbolic world without measure, the world that I have enhanced with the colors of my exaggeration, of my impulse, or of my whimsy – with all the entailed risks. A world built of untenable positions that places me morally above mundane pettiness. In the paradise that Cantor created for us, to which we are led by Hilbert, confronted with the world created by my fancy, the contortions of a mediocre self quenched with its rigor are totally irrelevant. If we are to persevere, if we are to prevail, we shall search for an incontestable and irrefutable reason, a Bachelardian reason that will yield only to our internal ghosts.

This is the lesson that some French thinkers learned when they learned the Austrian lesson, a lesson that does not reduce itself to the Vienna Circle, but of which the Circle is a crystal clear example. Because it was also at Vienna where we learned the lessons of Guardini and Freud.

Above all, towers the figure of Ludwig Wittgenstein. For a long time – too long – Wittgenstein appeared as the advocate of a philosophical position that we chose to fight on political grounds. To return to Wittgenstein and to discover that he is the most severe critic of a philosophical position with which he has been frequently associated has been one of the major contributions of French epistemology.

For when Wittgenstein says that the world is what is the case (*was der Fall ist*), he is focusing on a descent;[1] he is pointing to that place from whence we fell. He is referring to a defeat that endorses a kind of philosophy in which the most important component cannot be expressed, of which the substantial part is the effort we make while we try to say what cannot be said, our struggle to be near. After all, Bachelard is important because of his aesthetic determination of an object; an object that, according to Canguilhem, is opaque, undetermined and undeterminable, the object of the history of science; that which we see and live but we do not know. It is a form of thought striving toward the absolute which becomes, consequently, transcendent.

In trying to find Wittgenstein again, we stumble over the impenetrable object of thought, that which we know that is but that "is not there",[2] that which stands out as the most perfect harmony, as the most radiant beauty or as the most intense passion.

That is why the crisis of a Marxism that bound us to ignore Wittgenstein and the Circle, or at least to condemn them by way of a very narrow concept of reason, is not a cause for alarm or for theoretical distress but rather an opportunity to enlarge our notion of rationality: was it not Marx who claimed that man is a project and the human condition is an expectation? Marx surely gave us both, and most surely, when both collapsed, we confirmed that, for Marxism, nothing human is foreign, defeat and fall included. Marxism, while being apparently disqualified, shows its real human essence, Marxism is transformed, then, into a human form of thought, it becomes a murmur in search of that *locus imaginarius* that Kant posed beyond all possible experience.

We are thus, inaugurating a dialogue that will allow us to start again the search for our true face in the circular mirror of imagination and the passions of Carnap, Gödel and Wittgenstein.

We meet here to build a new understanding of our tradition, to redress our grievances with our own history.

Thus, we will come closer to our transcendental condition and to reopen the doors of the paradise that Cantor created for us and we will be capable of recovering the infinite tasks of thought that Gödel laid before us.

As Cavaillès, who learned from Bolzano and from Carnap, who heard Wittgenstein's silence and understood infinity from Gödel, we might be able to say, at last, "*Je suis accordé avec l'univers*".[3]

NOTES

* I wish to acknowledge the fruitful discussions with E. Koehler and Hao Wang and the continued and passionate effort of Carlos Torres to explain, time and time again, the subtleties of Gödel's theorems.

** This paper is a modified version of the one delivered at the International Seminar on Mathematical Philosophy, held in Oaxtepec, Mexico in 1991.

[1] The word 'case' comes from the Latin *casus* meaning 'fall'. Case is the past participate.

[2] The idea behind this characterization of being cannot be expressed in English where the difference between the Latin verbs *stare* and *essere* is lost. This ontological feature of Spanish is expressed in the very subtle distinction between the verbs *estar* and *ser*. The closest approximation could be rendered metaphorically as the difference between being that 'sits' (*essere*) and being that 'stands (*stare*). This is an essential difference to understand the concept of the Greek *physis* as that which stands by itself. The second substantial difference between the Spanish language and English is the fact that the verb 'born' is, in Spanish, an active verb, allowing expressions such as 'I born' or as 'nature borns' (*natura naturans*) that have no sense in English where the verb is always passive. This, as Hao Wang points out, generates a grammar (in the sense of Wittgenstein) that leads to the idea of God in expressions such as 'I gave birth to nature', etc.

[3] I have reached an agreement with the universe.

LIST OF CONTRIBUTORS

Carlos Alvarez Jimenez received his Ph.D. at the University of Amiens, France with a thesis on mathematical formalism and logicism. He is currently working on the history and the philosophy of the mathematics developed by the Warsaw school. He is a professor of mathematics at the Department of Mathematics of the National University of Mexico.

Ana Barahona Echeverría is a professor in the Department of Biology at the National University of Mexico. She obtained her Ph.D. from the National University of Mexico with a work on the History of Genetics. The work we present here is a résumé of her dissertation.

Walter Beller is a professor at the Metropolitan University in Mexico City. He is coauthor of *El positivismo mexicano*. He has worked on Genetic Epistemology. He obtained BA's in Law and Philosophy from the National University of Mexico.

Jaime Oscar Falcón Vega has a doctorate in physics at the University of Paris and was Professor of Mathematics at the National University of Mexico until 1985, after which he moved to Berlin where he has been working on chaos theory.

Max Fernández de Castro is a professor of logic at the Metropolitan University in Mexico City. His main interest concerns Husserl's philosophy of mathematics.

Alejandro R. Garciadiego obtained his Ph.D. at the University of Toronto. He is a member, together with Santiago Ramírez, of the editorial board of *Mathesis*. He is well known for his works on Bertrand Russell. He is a professor of the history of mathematics at the National University of Mexico.

Rafael Martínez E. obtained his MA at the Open University in England in 1978. He was head of the Department of Mathematics of UNAM in 1980–1981 and again in 1988–1989. The paper we offer here obtained the first prize in a contest organized by the Mexican Society of Physics to celebrate Newton's third centennial.

S. Ramirez and R. S. Cohen (eds.), Mexican Studies in the History and Philosophy of Science, 275–276.

Elia Nathan Bravo obtained her MA at Brandeis University and is associated with the Institute for Philosophical Research at the National University of Mexico.

Carlos Pereyra (1940–1988) was a professor of Political Philosophy and Philosophy of History at the National University of Mexico. He published several works on the problem of the subject of history. He was a remarkable political journalist. He died of cancer at the age of 48 in 1988.

Santiago Ramírez Castaneda has a Ph.D. from the University of Paris. He is a professor at the schools of Philosophy and of Science at the National University of Mexico and has been a visiting scholar at Harvard and Boston Universities.

Santiago Ramírez Ruiz (1921–1989) was the founder of psychoanalysis in Mexico. He was also a founder of the School of Psychology of the National University. His most famed work is *El Mexicano, psicología de sus motivaciones* that, together with Octavio Paz has become a classic in the search for a national identity. The paper we offer here was delivered at the First National Congress on Philosophy in 1975.

José Ruiz de Esparza is a researcher at the Institute for Historical Research. He was the editor of the works of Santiago Ramírez.

Rosaura Ruiz Gutierrez was the head of the Graduate Division of the Department of Biology at the National University of Mexico. She obtained her Ph.D. with a thesis about the introduction of Darwinism in Mexico which obtained the ADHILAC prize in 1984.

Carlos Torres is currently a professor of Mathematical Logic at the Department of Mathematics of the National University of Mexico, where he received his MA degree with a thesis on Hilbert's program.

Corina Yturbe is currently a member of the Institute for Philosophical Research at the National University of Mexico and a professor of Political Philosophy and Philosophy of History at the School of Philosophy.

Guillermo Zambrana Castañeda is currently a professor of the History of Mathematics at the National University of Mexico. He works on Wittgenstein's mathematical reflections and also on early Greek mathematics.

INDEX

S. Ramirez and R. S. Cohen (eds.), Mexican Studies in the History and Philosophy of Science, 277–280.

Boston Studies in the Philosophy of Science

Editor: Robert S. Cohen, *Boston University*

1. M.W. Wartofsky (ed.): *Proceedings of the Boston Colloquium for the Philosophy of Science, 1961/1962.* [Synthese Library 6] 1963
ISBN 90-277-0021-4
2. R.S. Cohen and M.W. Wartofsky (eds.): *Proceedings of the Boston Colloquium for the Philosophy of Science, 1962/1964.* In Honor of P. Frank. [Synthese Library 10] 1965 ISBN 90-277-9004-0
3. R.S. Cohen and M.W. Wartofsky (eds.): *Proceedings of the Boston Colloquium for the Philosophy of Science, 1964/1966.* In Memory of Norwood Russell Hanson. [Synthese Library 14] 1967 ISBN 90-277-0013-3
4. R.S. Cohen and M.W. Wartofsky (eds.): *Proceedings of the Boston Colloquium for the Philosophy of Science, 1966/1968.* [Synthese Library 18] 1969
ISBN 90-277-0014-1
5. R.S. Cohen and M.W. Wartofsky (eds.): *Proceedings of the Boston Colloquium for the Philosophy of Science, 1966/1968.* [Synthese Library 19] 1969
ISBN 90-277-0015-X
6. R.S. Cohen and R.J. Seeger (eds.): *Ernst Mach, Physicist and Philosopher.* [Synthese Library 27] 1970 ISBN 90-277-0016-8
7. M. Čapek: *Bergson and Modern Physics.* A Reinterpretation and Re-evaluation. [Synthese Library 37] 1971 ISBN 90-277-0186-5
8. R.C. Buck and R.S. Cohen (eds.): *PSA 1970.* Proceedings of the 2nd Biennial Meeting of the Philosophy and Science Association (Boston, Fall 1970). In Memory of Rudolf Carnap. [Synthese Library 39] 1971
ISBN 90-277-0187-3; Pb 90-277-0309-4
9. A.A. Zinov'ev: *Foundations of the Logical Theory of Scientific Knowledge (Complex Logic).* Translated from Russian. Revised and enlarged English Edition, with an Appendix by G.A. Smirnov, E.A. Sidorenko, A.M. Fedina and L.A. Bobrova. [Synthese Library 46] 1973
ISBN 90-277-0193-8; Pb 90-277-0324-8
10. L. Tondl: *Scientific Procedures.* A Contribution Concerning the Methodological Problems of Scientific Concepts and Scientific Explanation.Translated from Czech. [Synthese Library 47] 1973 ISBN 90-277-0147-4; Pb 90-277-0323-X
11. R.J. Seeger and R.S. Cohen (eds.): *Philosophical Foundations of Science.* Proceedings of Section L, 1969, American Association for the Advancement of Science. [Synthese Library 58] 1974 ISBN 90-277-0390-6; Pb 90-277-0376-0
12. A. Grünbaum: *Philosophical Problems of Space and Times.* 2nd enlarged ed. [Synthese Library 55] 1973 ISBN 90-277-0357-4; Pb 90-277-0358-2
13. R.S. Cohen and M.W. Wartofsky (eds.): *Logical and Epistemological Studies in Contemporary Physics.* Proceedings of the Boston Colloquium for the Philosophy of Science, 1969/72, Part I. [Synthese Library 59] 1974
ISBN 90-277-0391-4; Pb 90-277-0377-9

14. R.S. Cohen and M.W. Wartofsky (eds.): *Methodological and Historical Essays in the Natural and Social Sciences.* Proceedings of the Boston Colloquium for the Philosophy of Science, 1969/72, Part II. [Synthese Library 60] 1974 ISBN 90-277-0392-2; Pb 90-277-0378-7
15. R.S. Cohen, J.J. Stachel and M.W. Wartofsky (eds.): *For Dirk Struik.* Scientific, Historical and Political Essays in Honor of Dirk J. Struik. [Synthese Library 61] 1974 ISBN 90-277-0393-0; Pb 90-277-0379-5
16. N. Geschwind: *Selected Papers on Language and the Brains.* [Synthese Library 68] 1974 ISBN 90-277-0262-4; Pb 90-277-0263-2
17. B.G. Kuznetsov: *Reason and Being.* Translated from Russian. Edited by C.R. Fawcett and R.S. Cohen. 1987 ISBN 90-277-2181-5
18. P. Mittelstaedt: *Philosophical Problems of Modern Physics.* Translated from the revised 4th German edition by W. Riemer and edited by R.S. Cohen. [Synthese Library 95] 1976 ISBN 90-277-0285-3; Pb 90-277-0506-2
19. H. Mehlberg: *Time, Causality, and the Quantum Theory.* Studies in the Philosophy of Science. Vol. I: *Essay on the Causal Theory of Time.* Vol. II: *Time in a Quantized Universe.* Translated from French. Edited by R.S. Cohen. 1980 Vol. I: ISBN 90-277-0721-9; Pb 90-277-1074-0
    Vol. II: ISBN 90-277-1075-9; Pb 90-277-1076-7
20. K.F. Schaffner and R.S. Cohen (eds.): *PSA 1972.* Proceedings of the 3rd Biennial Meeting of the Philosophy of Science Association (Lansing, Michigan, Fall 1972). [Synthese Library 64] 1974 ISBN 90-277-0408-2; Pb 90-277-0409-0
21. R.S. Cohen and J.J. Stachel (eds.): *Selected Papers of Léon Rosenfeld.* [Synthese Library 100] 1979 ISBN 90-277-0651-4; Pb 90-277-0652-2
22. M. Čapek (ed.): *The Concepts of Space and Time.* Their Structure and Their Development. [Synthese Library 74] 1976 ISBN 90-277-0355-8; Pb 90-277-0375-2
23. M. Grene: *The Understanding of Nature.* Essays in the Philosophy of Biology. [Synthese Library 66] 1974 ISBN 90-277-0462-7; Pb 90-277-0463-5
24. D. Ihde: *Technics and Praxis.* A Philosophy of Technology. [Synthese Library 130] 1979 ISBN 90-277-0953-X; Pb 90-277-0954-8
25. J. Hintikka and U. Remes: *The Method of Analysis.* Its Geometrical Origin and Its General Significance. [Synthese Library 75] 1974 ISBN 90-277-0532-1; Pb 90-277-0543-7
26. J.E. Murdoch and E.D. Sylla (eds.): *The Cultural Context of Medieval Learning.* Proceedings of the First International Colloquium on Philosophy, Science, and Theology in the Middle Ages, 1973. [Synthese Library 76] 1975 ISBN 90-277-0560-7; Pb 90-277-0587-9
27. M. Grene and E. Mendelsohn (eds.): *Topics in the Philosophy of Biology.* [Synthese Library 84] 1976 ISBN 90-277-0595-X; Pb 90-277-0596-8
28. J. Agassi: *Science in Flux.* [Synthese Library 80] 1975 ISBN 90-277-0584-4; Pb 90-277-0612-3

29. J.J. Wiatr (ed.): *Polish Essays in the Methodology of the Social Sciences.* [Synthese Library 131] 1979 ISBN 90-277-0723-5; Pb 90-277-0956-4
30. P. Janich: *Protophysics of Time.* Constructive Foundation and History of Time Measurement. Translated from German. 1985 ISBN 90-277-0724-3
31. R.S. Cohen and M.W. Wartofsky (eds.): *Language, Logic, and Method.* 1983 ISBN 90-277-0725-1
32. R.S. Cohen, C.A. Hooker, A.C. Michalos and J.W. van Evra (eds.): *PSA 1974.* Proceedings of the 4th Biennial Meeting of the Philosophy of Science Association. [Synthese Library 101] 1976 ISBN 90-277-0647-6; Pb 90-277-0648-4
33. G. Holton and W.A. Blanpied (eds.): *Science and Its Public.* The Changing Relationship. [Synthese Library 96] 1976 ISBN 90-277-0657-3; Pb 90-277-0658-1
34. M.D. Grmek, R.S. Cohen and G. Cimino (eds.): *On Scientific Discovery.* The 1977 Erice Lectures. 1981 ISBN 90-277-1122-4; Pb 90-277-1123-2
35. S. Amsterdamski: *Between Experience and Metaphysics.* Philosophical Problems of the Evolution of Science. Translated from Polish. [Synthese Library 77] 1975 ISBN 90-277-0568-2; Pb 90-277-0580-1
36. M. Marković and G. Petrović (eds.): *Praxis.* Yugoslav Essays in the Philosophy and Methodology of the Social Sciences. [Synthese Library 134] 1979 ISBN 90-277-0727-8; Pb 90-277-0968-8
37. H. von Helmholtz: *Epistemological Writings.* The Paul Hertz / Moritz Schlick Centenary Edition of 1921. Translated from German by M.F. Lowe. Edited with an Introduction and Bibliography by R.S. Cohen and Y. Elkana. [Synthese Library 79] 1977 ISBN 90-277-0290-X; Pb 90-277-0582-8
38. R.M. Martin: *Pragmatics, Truth and Language.* 1979 ISBN 90-277-0992-0; Pb 90-277-0993-9
39. R.S. Cohen, P.K. Feyerabend and M.W. Wartofsky (eds.): *Essays in Memory of Imre Lakatos.* [Synthese Library 99] 1976 ISBN 90-277-0654-9; Pb 90-277-0655-7
40. Not published.
41. Not published.
42. H.R. Maturana and F.J. Varela: *Autopoiesis and Cognition.* The Realization of the Living. With a Preface to 'Autopoiesis' by S. Beer. 1980 ISBN 90-277-1015-5; Pb 90-277-1016-3
43. A. Kasher (ed.): *Language in Focus: Foundations, Methods and Systems.* Essays in Memory of Yehoshua Bar-Hillel. [Synthese Library 89] 1976 ISBN 90-277-0644-1; Pb 90-277-0645-X
44. T.D. Thao: *Investigations into the Origin of Language and Consciousness.* 1984 ISBN 90-277-0827-4
45. Not published.
46. P.L. Kapitza: *Experiment, Theory, Practice.* Articles and Addresses. Edited by R.S. Cohen. 1980 ISBN 90-277-1061-9; Pb 90-277-1062-7

Boston Studies in the Philosophy of Science

47. M.L. Dalla Chiara (ed.): *Italian Studies in the Philosophy of Science.* 1981
ISBN 90-277-0735-9; Pb 90-277-1073-2
48. M.W. Wartofsky: *Models.* Representation and the Scientific Understanding. [Synthese Library 129] 1979 ISBN 90-277-0736-7; Pb 90-277-0947-5
49. T.D. Thao: *Phenomenology and Dialectical Materialism.* Edited by R.S. Cohen. 1986 ISBN 90-277-0737-5
50. Y. Fried and J. Agassi: *Paranoia.* A Study in Diagnosis. [Synthese Library 102] 1976 ISBN 90-277-0704-9; Pb 90-277-0705-7
51. K.H. Wolff: *Surrender and Cath.* Experience and Inquiry Today. [Synthese Library 105] 1976 ISBN 90-277-0758-8; Pb 90-277-0765-0
52. K. Kosík: *Dialectics of the Concrete.* A Study on Problems of Man and World. 1976 ISBN 90-277-0761-8; Pb 90-277-0764-2
53. N. Goodman: *The Structure of Appearance.* [Synthese Library 107] 1977
ISBN 90-277-0773-1; Pb 90-277-0774-X
54. H.A. Simon: *Models of Discovery* and Other Topics in the Methods of Science. [Synthese Library 114] 1977 ISBN 90-277-0812-6; Pb 90-277-0858-4
55. M. Lazerowitz: *The Language of Philosophy.* Freud and Wittgenstein. [Synthese Library 117] 1977 ISBN 90-277-0826-6; Pb 90-277-0862-2
56. T. Nickles (ed.): *Scientific Discovery, Logic, and Rationality.* 1980
ISBN 90-277-1069-4; Pb 90-277-1070-8
57. J. Margolis: *Persons and Mind.* The Prospects of Nonreductive Materialism. [Synthese Library 121] 1978 ISBN 90-277-0854-1; Pb 90-277-0863-0
58. G. Radnitzky and G. Andersson (eds.): *Progress and Rationality in Science.* [Synthese Library 125] 1978 ISBN 90-277-0921-1; Pb 90-277-0922-X
59. G. Radnitzky and G. Andersson (eds.): *The Structure and Development of Science.* [Synthese Library 136] 1979 ISBN 90-277-0994-7; Pb 90-277-0995-5
60. T. Nickles (ed.): *Scientific Discovery.* Case Studies. 1980
ISBN 90-277-1092-9; Pb 90-277-1093-7
61. M.A. Finocchiaro: *Galileo and the Art of Reasoning.* Rhetorical Foundation of Logic and Scientific Method. 1980 ISBN 90-277-1094-5; Pb 90-277-1095-3
62. W.A. Wallace: *Prelude to Galileo.* Essays on Medieval and 16th-Century Sources of Galileo's Thought. 1981 ISBN 90-277-1215-8; Pb 90-277-1216-6
63. F. Rapp: *Analytical Philosophy of Technology.* Translated from German. 1981
ISBN 90-277-1221-2; Pb 90-277-1222-0
64. R.S. Cohen and M.W. Wartofsky (eds.): *Hegel and the Sciences.* 1984
ISBN 90-277-0726-X
65. J. Agassi: *Science and Society.* Studies in the Sociology of Science. 1981
ISBN 90-277-1244-1; Pb 90-277-1245-X
66. L. Tondl: *Problems of Semantics.* A Contribution to the Analysis of the Language of Science. Translated from Czech. 1981
ISBN 90-277-0148-2; Pb 90-277-0316-7
67. J. Agassi and R.S. Cohen (eds.): *Scientific Philosophy Today.* Essays in Honor of Mario Bunge. 1982 ISBN 90-277-1262-X; Pb 90-277-1263-8

Boston Studies in the Philosophy of Science

68. W. Krajewski (ed.): *Polish Essays in the Philosophy of the Natural Sciences.* Translated from Polish and edited by R.S. Cohen and C.R. Fawcett. 1982 ISBN 90-277-1286-7; Pb 90-277-1287-5
69. J.H. Fetzer: *Scientific Knowledge.* Causation, Explanation and Corroboration. 1981 ISBN 90-277-1335-9; Pb 90-277-1336-7
70. S. Grossberg: *Studies of Mind and Brain.* Neural Principles of Learning, Perception, Development, Cognition, and Motor Control. 1982 ISBN 90-277-1359-6; Pb 90-277-1360-X
71. R.S. Cohen and M.W. Wartofsky (eds.): *Epistemology, Methodology, and the Social Sciences.* 1983. ISBN 90-277-1454-1
72. K. Berka: *Measurement.* Its Concepts, Theories and Problems. Translated from Czech. 1983 ISBN 90-277-1416-9
73. G.L. Pandit: *The Structure and Growth of Scientific Knowledge.* A Study in the Methodology of Epistemic Appraisal. 1983 ISBN 90-277-1434-7
74. A.A. Zinov'ev: *Logical Physics.* Translated from Russian. Edited by R.S. Cohen. 1983 ISBN 90-277-0734-0
    See also Volume 9.
75. G-G. Granger: *Formal Thought and the Sciences of Man.* Translated from French. With and Introduction by A. Rosenberg. 1983 ISBN 90-277-1524-6
76. R.S. Cohen and L. Laudan (eds.): *Physics, Philosophy and Psychoanalysis.* Essays in Honor of Adolf Grünbaum. 1983 ISBN 90-277-1533-5
77. G. Böhme, W. van den Daele, R. Hohlfeld, W. Krohn and W. Schäfer: *Finalization in Science.* The Social Orientation of Scientific Progress. Translated from German. Edited by W. Schäfer. 1983 ISBN 90-277-1549-1
78. D. Shapere: *Reason and the Search for Knowledge.* Investigations in the Philosophy of Science. 1984 ISBN 90-277-1551-3; Pb 90-277-1641-2
79. G. Andersson (ed.): *Rationality in Science and Politics.* Translated from German. 1984 ISBN 90-277-1575-0; Pb 90-277-1953-5
80. P.T. Durbin and F. Rapp (eds.): *Philosophy and Technology.* [*Also* Philosophy and Technology Series, Vol. 1] 1983 ISBN 90-277-1576-9
81. M. Marković: *Dialectical Theory of Meaning.* Translated from Serbo-Croat. 1984 ISBN 90-277-1596-3
82. R.S. Cohen and M.W. Wartofsky (eds.): *Physical Sciences and History of Physics.* 1984. ISBN 90-277-1615-3
83. É. Meyerson: *The Relativistic Deduction.* Epistemological Implications of the Theory of Relativity. Translated from French. With a Review by Albert Einstein and an Introduction by Milić Čapek. 1985 ISBN 90-277-1699-4
84. R.S. Cohen and M.W. Wartofsky (eds.): *Methodology, Metaphysics and the History of Science.* In Memory of Benjamin Nelson. 1984 ISBN 90-277-1711-7
85. G. Tamás: *The Logic of Categories.* Translated from Hungarian. Edited by R.S. Cohen. 1986 ISBN 90-277-1742-7
86. S.L. de C. Fernandes: *Foundations of Objective Knowledge.* The Relations of Popper's Theory of Knowledge to That of Kant. 1985 ISBN 90-277-1809-1

87. R.S. Cohen and T. Schnelle (eds.): *Cognition and Fact.* Materials on Ludwik Fleck. 1986 ISBN 90-277-1902-0
88. G. Freudenthal: *Atom and Individual in the Age of Newton.* On the Genesis of the Mechanistic World View. Translated from German. 1986 ISBN 90-277-1905-5
89. A. Donagan, A.N. Perovich Jr and M.V. Wedin (eds.): *Human Nature and Natural Knowledge.* Essays presented to Marjorie Grene on the Occasion of Her 75th Birthday. 1986 ISBN 90-277-1974-8
90. C. Mitcham and A. Hunning (eds.): *Philosophy and Technology II.* Information Technology and Computers in Theory and Practice. [*Also* Philosophy and Technology Series, Vol. 2] 1986 ISBN 90-277-1975-6
91. M. Grene and D. Nails (eds.): *Spinoza and the Sciences.* 1986 ISBN 90-277-1976-4
92. S.P. Turner: *The Search for a Methodology of Social Science.* Durkheim, Weber, and the 19th-Century Problem of Cause, Probability, and Action. 1986. ISBN 90-277-2067-3
93. I.C. Jarvie: *Thinking about Society.* Theory and Practice. 1986 ISBN 90-277-2068-1
94. E. Ullmann-Margalit (ed.): *The Kaleidoscope of Science.* The Israel Colloquium: Studies in History, Philosophy, and Sociology of Science, Vol. 1. 1986 ISBN 90-277-2158-0; Pb 90-277-2159-9
95. E. Ullmann-Margalit (ed.): *The Prism of Science.* The Israel Colloquium: Studies in History, Philosophy, and Sociology of Science, Vol. 2. 1986 ISBN 90-277-2160-2; Pb 90-277-2161-0
96. G. Márkus: *Language and Production.* A Critique of the Paradigms. Translated from French. 1986 ISBN 90-277-2169-6
97. F. Amrine, F.J. Zucker and H. Wheeler (eds.): *Goethe and the Sciences: A Reappraisal.* 1987 ISBN 90-277-2265-X; Pb 90-277-2400-8
98. J.C. Pitt and M. Pera (eds.): *Rational Changes in Science.* Essays on Scientific Reasoning. Translated from Italian. 1987 ISBN 90-277-2417-2
99. O. Costa de Beauregard: *Time, the Physical Magnitude.* 1987 ISBN 90-277-2444-X
100. A. Shimony and D. Nails (eds.): *Naturalistic Epistemology.* A Symposium of Two Decades. 1987 ISBN 90-277-2337-0
101. N. Rotenstreich: *Time and Meaning in History.* 1987 ISBN 90-277-2467-9
102. D.B. Zilberman: *The Birth of Meaning in Hindu Thought.* Edited by R.S. Cohen. 1988 ISBN 90-277-2497-0
103. T.F. Glick (ed.): *The Comparative Reception of Relativity.* 1987 ISBN 90-277-2498-9
104. Z. Harris, M. Gottfried, T. Ryckman, P. Mattick Jr, A. Daladier, T.N. Harris and S. Harris: *The Form of Information in Science.* Analysis of an Immunology Sublanguage. With a Preface by Hilary Putnam. 1989 ISBN 90-277-2516-0

Boston Studies in the Philosophy of Science

105. F. Burwick (ed.): *Approaches to Organic Form.* Permutations in Science and Culture. 1987 ISBN 90-277-2541-1
106. M. Almási: *The Philosophy of Appearances.* Translated from Hungarian. 1989 ISBN 90-277-2150-5
107. S. Hook, W.L. O'Neill and R. O'Toole (eds.): *Philosophy, History and Social Action.* Essays in Honor of Lewis Feuer. With an Autobiographical Essay by L. Feuer. 1988 ISBN 90-277-2644-2
108. I. Hronszky, M. Fehér and B. Dajka: *Scientific Knowledge Socialized.* Selected Proceedings of the 5th Joint International Conference on the History and Philosophy of Science organized by the IUHPS (Veszprém, Hungary, 1984). 1988 ISBN 90-277-2284-6
109. P. Tillers and E.D. Green (eds.): *Probability and Inference in the Law of Evidence.* The Uses and Limits of Bayesianism. 1988 ISBN 90-277-2689-2
110. E. Ullmann-Margalit (ed.): *Science in Reflection.* The Israel Colloquium: Studies in History, Philosophy, and Sociology of Science, Vol. 3. 1988 ISBN 90-277-2712-0; Pb 90-277-2713-9
111. K. Gavroglu, Y. Goudaroulis and P. Nicolacopoulos (eds.): *Imre Lakatos and Theories of Scientific Change.* 1989 ISBN 90-277-2766-X
112. B. Glassner and J.D. Moreno (eds.): *The Qualitative-Quantitative Distinction in the Social Sciences.* 1989 ISBN 90-277-2829-1
113. K. Arens: *Structures of Knowing.* Psychologies of the 19th Century. 1989 ISBN 0-7923-0009-2
114. A. Janik: *Style, Politics and the Future of Philosophy.* 1989 ISBN 0-7923-0056-4
115. F. Amrine (ed.): *Literature and Science as Modes of Expression.* With an Introduction by S. Weininger. 1989 ISBN 0-7923-0133-1
116. J.R. Brown and J. Mittelstrass (eds.): *An Intimate Relation.* Studies in the History and Philosophy of Science. Presented to Robert E. Butts on His 60th Birthday. 1989 ISBN 0-7923-0169-2
117. F. D'Agostino and I.C. Jarvie (eds.): *Freedom and Rationality.* Essays in Honor of John Watkins. 1989 ISBN 0-7923-0264-8
118. D. Zolo: *Reflexive Epistemology.* The Philosophical Legacy of Otto Neurath. 1989 ISBN 0-7923-0320-2
119. M. Kearn, B.S. Philips and R.S. Cohen (eds.): *Georg Simmel and Contemporary Sociology.* 1989 ISBN 0-7923-0407-1
120. T.H. Levere and W.R. Shea (eds.): *Nature, Experiment and the Science.* Essays on Galileo and the Nature of Science. In Honour of Stillman Drake. 1989 ISBN 0-7923-0420-9
121. P. Nicolacopoulos (ed.): *Greek Studies in the Philosophy and History of Science.* 1990 ISBN 0-7923-0717-8
122. R. Cooke and D. Costantini (eds.): *Statistics in Science.* The Foundations of Statistical Methods in Biology, Physics and Economics. 1990 ISBN 0-7923-0797-6

Boston Studies in the Philosophy of Science

123. P. Duhem: *The Origins of Statics*. Translated from French by G.F. Leneaux, V.N. Vagliente and G.H. Wagner. With an Introduction by S.L. Jaki. 1991 ISBN 0-7923-0898-0
124. H. Kamerlingh Onnes: *Through Measurement to Knowledge*. The Selected Papers, 1853-1926. Edited and with an Introduction by K. Gavroglu and Y. Goudaroulis. 1991 ISBN 0-7923-0825-5
125. M. Čapek: *The New Aspects of Time: Its Continuity and Novelties*. Selected Papers in the Philosophy of Science. 1991 ISBN 0-7923-0911-1
126. S. Unguru (ed.): *Physics, Cosmology and Astronomy, 1300-1700*. Tension and Accommodation. 1991 ISBN 0-7923-1022-5
127. Z. Bechler: *Newton's Physics on the Conceptual Structure of the Scientific Revolution*. 1991 ISBN 0-7923-1054-3
128. É. Meyerson: *Explanation in the Sciences*. Translated from French by M-A. Siple and D.A. Siple. 1991 ISBN 0-7923-1129-9
129. A.I. Tauber (ed.): *Organism and the Origins of Self*. 1991 ISBN 0-7923-1185-X
130. F.J. Varela and J-P. Dupuy (eds.): *Understanding Origins*. Contemporary Views on the Origin of Life, Mind and Society. 1992 ISBN 0-7923-1251-1
131. G.L. Pandit: *Methodological Variance*. Essays in Epistemological Ontology and the Methodology of Science. 1991 ISBN 0-7923-1263-5
132. G. Munévar (ed.): *Beyond Reason*. Essays on the Philosophy of Paul Feyerabend. 1991 ISBN 0-7923-1272-4
133. T.E. Uebel (ed.): *Rediscovering the Forgotten Vienna Circle*. Austrian Studies on Otto Neurath and the Vienna Circle. Partly translated from German. 1991 ISBN 0-7923-1276-7
134. W.R. Woodward and R.S. Cohen (eds.): *World Views and Scientific Discipline Formation*. Science Studies in the [former] German Democratic Republic. Partly translated from German by W.R. Woodward. 1991 ISBN 0-7923-1286-4
135. P. Zambelli: *The Speculum Astronomiae and Its Enigma*. Astrology, Theology and Science in Albertus Magnus and His Contemporaries. 1992 ISBN 0-7923-1380-1
136. P. Petitjean, C. Jami and A.M. Moulin (eds.): *Science and Empires*. Historical Studies about Scientific Development and European Expansion. ISBN 0-7923-1518-9
137. W.A. Wallace: *Galileo's Logic of Discovery and Proof*. The Background, Content, and Use of His Appropriated Treatises on Aristotle's *Posterior Analytics*. 1992 ISBN 0-7923-1577-4
138. W.A. Wallace: *Galileo's Logical Treatises*. A Translation, with Notes and Commentary, of His Appropriated Latin Questions on Aristotle's *Posterior Analytics*. 1992 ISBN 0-7923-1578-2
     Set (137 + 138) ISBN 0-7923-1579-0

Boston Studies in the Philosophy of Science

139. M.J. Nye, J.L. Richards and R.H. Stuewer (eds.): *The Invention of Physical Science.* Intersections of Mathematics, Theology and Natural Philosophy since the Seventeenth Century. Essays in Honor of Erwin N. Hiebert. 1992 ISBN 0-7923-1753-X
140. G. Corsi, M.L. dalla Chiara and G.C. Ghirardi (eds.): *Bridging the Gap: Philosophy, Mathematics and Physics.* Lectures on the Foundations of Science. 1992 ISBN 0-7923-1761-0
141. C.-H. Lin and D. Fu (eds.): *Philosophy and Conceptual History of Science in Taiwan.* 1992 ISBN 0-7923-1766-1
142. S. Sarkar (ed.): *The Founders of Evolutionary Genetics.* A Centenary Reappraisal. 1992 ISBN 0-7923-1777-7
143. J. Blackmore (ed.): *Ernst Mach – A Deeper Look.* Documents and New Perspectives. 1992 ISBN 0-7923-1853-6
144. P. Kroes and M. Bakker (eds.): *Technological Development and Science in the Industrial Age.* New Perspectives on the Science–Technology Relationship. 1992 ISBN 0-7923-1898-6
145. S. Amsterdamski: *Between History and Method.* Disputes about the Rationality of Science. 1992 ISBN 0-7923-1941-9
146. E. Ullmann-Margalit (ed.): *The Scientific Enterprise.* The Bar-Hillel Colloquium: Studies in History, Philosophy, and Sociology of Science, Volume 4. 1992 ISBN 0-7923-1992-3
147. L. Embree (ed.): *Metaarchaeology.* Reflections by Archaeologists and Philosophers. 1992 ISBN 0-7923-2023-9
148. S. French and H. Kamminga (eds.): *Correspondence, Invariance and Heuristics.* Essays in Honour of Heinz Post. 1993 ISBN 0-7923-2085-9
149. M. Bunzl: *The Context of Explanation.* 1993 ISBN 0-7923-2153-7
150. I.B. Cohen (ed.): *The Natural Sciences and the Social Sciences.* Some Critical and Historical Perspectives. 1994 ISBN 0-7923-2223-1
151. K. Gavroglu, Y. Christianidis and E. Nicolaidis (eds.): *Trends in the Historiography of Science.* 1994 ISBN 0-7923-2255-X
152. S. Poggi and M. Bossi (eds.): *Romanticism in Science.* Science in Europe, 1790–1840. 1994 ISBN 0-7923-2336-X
153. J. Faye and H.J. Folse (eds.): *Niels Bohr and Contemporary Philosophy.* 1994 ISBN 0-7923-2378-5
154. C.C. Gould and R.S. Cohen (eds.): *Artifacts, Representations, and Social Practice.* Essays for Marx W. Wartofsky. 1994 ISBN 0-7923-2481-1
155. R.E. Butts: *Historical Pragmatics.* Philosophical Essays. 1993 ISBN 0-7923-2498-6
156. R. Rashed: *The Development of Arabic Mathematics: Between Arithmetic and Algebra.* Translated from French by A.F.W. Armstrong. 1994 ISBN 0-7923-2565-6

157. I. Szumilewicz-Lachman (ed.): *Zygmunt Zawirski: His Life and Work.* With Selected Writings on Time, Logic and the Methodology of Science. Translations by Feliks Lachman. Ed. by R.S. Cohen, with the assistance of B. Bergo. 1994 ISBN 0-7923-2566-4
158. S.N. Haq: *Names, Natures and Things.* The Alchemist Jābir ibn Ḥayyān and His *Kitāb al-Aḥjār* (Book of Stones). 1994 ISBN 0-7923-2587-7
159. P. Plaass: *Kant's Theory of Natural Science.* Translation, Analytic Introduction and Commentary by Alfred E. and Maria G. Miller. 1994 ISBN 0-7923-2750-0
160. J. Misiek (ed.): *The Problem of Rationality in Science and its Philosophy.* On Popper vs. Polanyi. The Polish Conferences 1988–89. 1995 ISBN 0-7923-2925-2
161. I.C. Jarvie and N. Laor (eds.): *Critical Rationalism, Metaphysics and Science.* Essays for Joseph Agassi, Volume I. 1995 ISBN 0-7923-2960-0
162. I.C. Jarvie and N. Laor (eds.): *Critical Rationalism, the Social Sciences and the Humanities.* Essays for Joseph Agassi, Volume II. 1995 ISBN 0-7923-2961-9
Set (161–162) ISBN 0-7923-2962-7
163. K. Gavroglu, J. Stachel and M.W. Wartofsky (eds.): *Physics, Philosophy, and the Scientific Community.* Essays in the Philosophy and History of the Natural Sciences and Mathematics. In Honor of Robert S. Cohen. 1995 ISBN 0-7923-2988-0
164. K. Gavroglu, J. Stachel and M.W. Wartofsky (eds.): *Science, Politics and Social Practice.* Essays on Marxism and Science, Philosophy of Culture and the Social Sciences. In Honor of Robert S. Cohen. 1995 ISBN 0-7923-2989-9
165. K. Gavroglu, J. Stachel and M.W. Wartofsky (eds.): *Science, Mind and Art.* Essays on Science and the Humanistic Understanding in Art, Epistemology, Religion and Ethics. Essays in Honor of Robert S. Cohen. 1995 ISBN 0-7923-2990-2
Set (163–165) ISBN 0-7923-2991-0
166. K.H. Wolff: *Transformation in the Writing.* A Case of Surrender-and-Catch. 1995 ISBN 0-7923-3178-8
167. A.J. Kox and D.M. Siegel (eds.): *No Truth Except in the Details.* Essays in Honor of Martin J. Klein. 1995 ISBN 0-7923-3195-8
168. J. Blackmore: *Ludwig Boltzmann, His Later Life and Philosophy, 1900–1906.* Book One: A Documentary History. 1995 ISBN 0-7923-3231-8
169. R.S. Cohen, R. Hilpinen and Q. Renzong (eds.): *Realism and Anti-Realism in the Philosophy of Science.* Beijing International Conference, 1992. 1995 ISBN 0-7923-3233-4
170. I. Kuçuradi and R.S. Cohen (eds.): *The Concept of Knowledge.* The Ankara Seminar. 1995 ISBN 0-7923-3241-5

Boston Studies in the Philosophy of Science

171. M.A. Grodin (ed.): *Meta Medical Ethics*: The Philosophical Foundations of Bioethics. 1995 ISBN 0-7923-3344-6
172. S. Ramirez and R.S. Cohen (eds.): *Mexican Studies in the History and Philosophy of Science*. 1995 ISBN 0-7923-3462-0
173. C. Dilworth: *The Metaphysics of Science*. An Account of Modern Science in Terms of Principles, Laws and Theories. 1995 ISBN 0-7923-3693-3
174. J. Blackmore: *Ludwig Boltzmann, His Later Life and Philosophy, 1900–1906* Book Two: The Philosopher. 1995 ISBN 0-7923-3464-7
175. P. Damerow: *Abstraction and Representation*. Essays on the Cultural Evolution of Thinking. 1995 ISBN 0-7923-3813-8
176. G. Tarozzi (ed.): *Karl Popper, Philosopher of Science*. (in prep.)
177. M. Marion and R.S. Cohen (eds.): *Québec Studies in the Philosophy of Science*. Part I: Logic, Mathematics, Physics and History of Science. Essays in Honor of Hugues Leblanc. 1995 ISBN 0-7923-3559-7
178. M. Marion and R.S. Cohen (eds.): *Québec Studies in the Philosophy of Science*. Part II: Biology, Psychology, Cognitive Science and Economics. Essays in Honor of Hugues Leblanc. 1996 ISBN 0-7923-3560-0
Set (177–178) ISBN 0-7923-3561-9
179. F. Dainian and R.S. Cohen (eds.): *Chinese Studies in the History and Philosophy of Science and Technology*. 1995 ISBN 0-7923-3463-9
180. P. Forman and J.M. Sánchez-Ron (eds.): *National Military Establishments and the Advancement of Science and Technology*. Studies in 20th Century History. 1995 ISBN 0-7923-3541-4
181. E.J. Post: *Quantum Reprogramming*. Ensembles and Single Systems: A Two-Tier Approach to Quantum Mechanics. 1995 ISBN 0-7923-3565-1

Also of interest:

R.S. Cohen and M.W. Wartofsky (eds.): *A Portrait of Twenty-Five Years Boston Colloquia for the Philosophy of Science, 1960-1985*. 1985 ISBN Pb 90-277-1971-3

Previous volumes are still available.

KLUWER ACADEMIC PUBLISHERS – DORDRECHT / BOSTON / LONDON

GPSR Compliance

The European Union's (EU) General Product Safety Regulation (GPSR) is a set of rules that requires consumer products to be safe and our obligations to ensure this.

If you have any concerns about our products, you can contact us on ProductSafety@springernature.com

In case Publisher is established outside the EU, the EU authorized representative is:

Springer Nature Customer Service Center GmbH
Europaplatz 3
69115 Heidelberg, Germany

Batch number: 10823846

Printed by Printforce, the Netherlands